SpringerBriefs in Computer Science

T0185336

For further volumes:
http://www.springer.com/series/10028

Wei Song · Weihua Zhuang

Interworking of Wireless LANs and Cellular Networks

 Springer

Wei Song
University of New Brunswick
Fredericton, NB
Canada

Weihua Zhuang
University of Waterloo
Waterloo, ON
Canada

ISSN 2191-5768   ISSN 2191-5776 (electronic)
ISBN 978-1-4614-4378-0   ISBN 978-1-4614-4379-7 (eBook)
DOI 10.1007/978-1-4614-4379-7
Springer New York Heidelberg Dordrecht London

Library of Congress Control Number: 2012940946

Printed on acid-free paper

Springer is part of Springer Science+Business Media (www.springer.com)

Preface

Next-generation wireless systems have been envisioned to be supported by heterogeneous wireless access technologies. The popular cellular networks and wireless local area networks (WLANs) present perfectly complementary characteristics in terms of service capacity, mobility support, and quality of service (QoS) provisioning. An essential aspect of the interworking is call admission control with effective access selection to achieve efficient resource utilization, QoS assurance, and load balancing. An incoming call needs to choose the overlay cell or WLAN according to an access selection strategy and follows an admission control policy in the target network. In this book, we investigate three interworking schemes for the cellular/WLAN integrated network.

In Chap. 2, we present an in-depth analysis of a simple *WLAN-first* resource allocation scheme, in which WLANs are always preferred whenever the WLAN access is available, so as to take advantage of the low cost and large bandwidth of WLANs. The analysis reveals important insights on the impact of admission regions. In Chap. 3, we introduce an admission scheme with randomized access selection to enable distributed implementation. Based on an analytical approach with moment generating functions (MGFs), we examine the impact of user mobility and data traffic variability on overall QoS satisfaction and resource utilization. In Chap. 4, a size-based load sharing scheme further takes into account heavy-tailed data call size to enhance QoS provisioning. Dynamic vertical handoff is considered to pool the available bandwidths of the two systems to improve the multiplexing gain.

This book presents an overview of the state-of-the-art solutions to cellular/ WLAN interworking. It not only reveals important observations but also offers useful tools for performance evaluation. The unique traffic and network characteristics are exploited to enhance interworking effectiveness. Theoretical analysis and simulation validation demonstrate benefits of cellular/ WLAN interworking in real networks. Last but not the least, this book highlights promising future research directions to guide interested readers.

Contents

Chapter 1
Introduction on Cellular/WLAN Interworking

Motivated by the ever-increasing demand for wireless communication services, the past decades have witnessed rapid evolution and successful deployment of wireless networks. It is widely accepted that next-generation wireless networks will be heterogeneous in nature with multiple wireless access technologies. While the heterogeneity poses new challenges to achieve inter-operability among different wireless networks, their complementary characteristics can be exploited with the interworking to enhance service provisioning. The popular cellular networks and wireless local area networks (WLANs) are two most promising radio access technologies. The cellular/WLAN interworking has attracted intensive research attention. In this chapter, we provide an overview of related work on cellular/WLAN interworking and an effective system model to capture the essential characteristics of cellular/WLAN integrated networks.

1.1 Interworking Research Issues

In the literature, many studies address the cellular/WLAN interworking issues such as vertical handoff, access selection, and call admission control.

1.1.1 Vertical Handoff

While the cellular network provides ubiquitous connectivity with wide-area coverage, WLANs are only deployed disjointly in hotspot areas. The cellular/WLAN interworking then results in an overlay structure, which offers both cellular access and WLAN access to dual-mode mobiles in WLAN-covered areas. The handoff between wireless networks of different access technologies is referred to as *vertical handoff*, in contrast to horizontal handoff within a homogeneous wireless network,

W. Song and W. Zhuang, *Interworking of Wireless LANs and Cellular Networks*,
SpringerBriefs in Computer Science, DOI: 10.1007/978-1-4614-4379-7_1,
© The Author(s) 2012

e.g., between base stations of cellular networks or access points of WLANs. It is necessary to differentiate the *downward vertical handoff* from a cell to a WLAN and *upward vertical handoff* from a WLAN to a cell. Further, vertical handoff may originate from quality-of-service (QoS) enhancement or load balancing considerations other than maintaining connectivity. Hence, not only can vertical handoff proceed when a mobile moves out of the cell/WLAN border, but also back-and-forth vertical handoff can take place when a mobile moves within the cell/WLAN.

Many vertical handoff algorithms are proposed to achieve seamless and fast handoff between the cell and the WLAN. Handoff decisions can be based on metrics such as received signal strength (RSS), signal-to-noise ratio (SNR), and user moving speed. Due to network heterogeneity, such traditional metrics in the two networks are rather distinct and should be used in a way different from that for horizontal handoff. The handoff algorithm proposed in [31] uses the number of continuous WLAN beacon signals whose strength falls below a predefined level. The handoff thresholds are differentiated according to handoff direction and traffic delay-sensitivity. In [56], two handoff decision algorithms are compared with respect to a user satisfaction function. The first algorithm requires handoff to the WLAN whenever it becomes available, while no handoff is allowed in the second algorithm if the mobile is engaged in real-time or streaming sessions. It is observed that the second algorithm outperforms the first for both indoor office scenarios and on-the-move mobility models.

Moreover, there are advanced vertical handoff decision algorithms, which simultaneously consider various factors such as network characteristics, service type, user mobility, network conditions, user preference, and cost. The handoff decision problem is first formulated to incorporate these factors and define the objectives, e.g., satisfaction of user requirement and maximization of revenue. Then, the decision problem needs to be solved efficiently with techniques such as fuzzy logic. In [50], the handoff decision is formulated as a fuzzy multiple attribute decision making (MADM) problem. Another vertical handoff approach is based on a Markov decision process (MDP) to determine the conditions to perform vertical handoff [48, 51, 52]. A link reward function and a signaling cost function are defined in the MDP formulation to capture the tradeoff between the network resources utilized by the connection and the signaling and processing load incurred on the network.

In addition to network-layer approaches, there are transport-layer and application-layer solutions. The transport-layer scheme proposed in [25] supports universal mobile telecommunications system (UMTS) and WLAN vertical handoff via stream control transmission protocol (SCTP). Although mobility management at the transport layer enables network-independence, more functions need to be introduced at end systems. A typical application-layer handoff solution is based on session initiation protocol (SIP) [4, 61]. SIP is a key signaling protocol for IP multimedia subsystem (IMS) of UMTS, in which real-time multimedia services are supported within a packet-switched domain. The UMTS-WLAN vertical handoff can take advantage of SIP to facilitate handoff-associated negotiation for QoS, authentication, authorization, and accounting (AAA). In general, application-layer solutions involve less modification to existing protocols and infrastructures of the cellular network and

WLANs. Nonetheless, relatively longer handoff latency may be incurred with the lower-layer network attachment and SIP location update.

1.1.2 Access Selection and Call Admission Control

With the cellular/WLAN interworking, there is ubiquitous cellular coverage, while both cellular access and WLAN access are available in the overlay area. Initially, an incoming new call should properly select either the covering cell or WLAN for access. The selected target network decides whether to accept or reject the call based on its admission control policy. If there is no sufficient available bandwidth to admit the call in the preferred network, the call can overflow to the other network or just leave the system. Moreover, if enough resources are released from call completion or outgoing handoff in the preferred network, an overflow call can reselect its preferred network. The access reselection is referred to as *take-back* in some literature.

Complementing the aforementioned access selection/reselection strategies, admission control policies in the target networks need to be properly designed to limit the admissible traffic load and provide QoS assurance. For example, as handoff dropping is more undesirable than new call blocking, handoff calls should be prioritized over new calls in the admission control policy, e.g., by means of reserving guard channels, queueing handoff calls, and so on. In addition, the admission control policy of wireless overlay networks needs to differentiate calls in different areas, since the accessible resources vary with locations.

The access selection and call admission control are fundamental functions of the network layer. Due to heterogeneous wireless access technologies at the link layer of the cellular network and WLANs, it is imperative to adapt these control functions to the available wireless links. Particularly, with the cellular/WLAN interworking, an incoming call should be properly routed and admitted to an underlying integrated system so that the requested QoS is supported efficiently. Different from wired networks, the wireless link capacity is highly time-varying and location-dependent. Hence, effective capacity evaluation is essential for admission decision. To ensure sufficient link capacity for QoS satisfaction, only when the admission request is accepted by the underlying link layer can the incoming call be admitted to a specific network.

The basic principles for allocating multiple services to multi-access wireless systems are discussed in [15]. Near-optimum subsystem service allocations that maximize combined capacity are derived through simple optimization procedures. The achievable capacity gain is found to be dependent on the characteristics of the individual subsystem capacity regions. Favorable service allocations are either extreme points where services are allocated to the subsystems best at supporting them, or characterized by equal relative efficiency for supporting services in all subsystems. Authors of [17] propose a Markovian approach for radio access selection in heterogeneous multiaccess/multiservice wireless networks. The proposed analytical framework addresses the allocation of two service types onto two access networks, which can be used to model a wide range of access selection policies taking into account several criteria such as service type, network conditions, and terminal types.

Similar to vertical handoff decision, multiple factors can be introduced to formulate the access selection problem and define the design objectives. Then, different techniques can be applied to rank the candidate access options. The decision process in [6] uses non-compensatory and compensatory multi-attribute decision making (MADM) algorithms jointly on the network side to assist the terminal to select the top candidate network. There are different MADM solutions such as multiplicative exponent weighting [59], simple additive weighting [62], and the technique for order preference by similarity to ideal solution (TOPSIS) [62]. The access selection in [42] is based on an integrated algorithm of analytic hierarchy process (AHP) and grey relational analysis (GRA). AHP quantitatively weights decision alternatives by hierarchical and pairwise comparison, while GRA ranks network alternatives efficiently through building a grey relationship with an ideal option.

In [41], multimedia application-layer QoS (e.g., video distortion) is considered as design criteria for access selection, which is different from the focus of network-layer QoS, such as blocking probability, throughput, and utilization. An optimal distributed network selection scheme is developed based on stochastic optimization.

Complementing access selection, call admission control is an essential network function for heterogeneous wireless networks. An analytical model is developed in [49] to facilitate the evaluation of different admission control policies in a multi-service integrated cellular/WLAN system. The system performance is evaluated for different combinations of cutoff priority [11] and fractional guard channel policies [34]. Game theory is another popular approach to investigate connection-level bandwidth allocation and admission control for heterogeneous wireless networks. It is observed in [30] that capacity reservation combined with network-level allocation can be used for vertical and horizontal handoff so that connection blocking and dropping probabilities are upper bounded. In [60], an optimal joint session admission control scheme is proposed for multimedia traffic. It is based on a semi-Markov decision process (SMDP) to maximize the overall network revenue with QoS constraints. Saturated traffic (i.e., there is always backlogged data in the service queue of an active session) is considered for all service classes, whose QoS requirements are differentiated by packet delay, saturation bandwidth, and signal-to-interference ratio (SIR).

1.2 System Modelling

As WLANs operate at license-exempt frequency bands, a large bandwidth is available to support a high date rate, e.g., up to 54 Mbit/s in IEEE 802.11a/g. In contrast, current widely deployed cellular networks support a relatively low data rate. For example, the UMTS system can provide a data rate up to 2 Mbit/s for low-mobility applications (up to 10 km/h) [19]. There are also some enhancement technologies such as the high speed packet access (HSPA), which can promote the downlink packet rate up

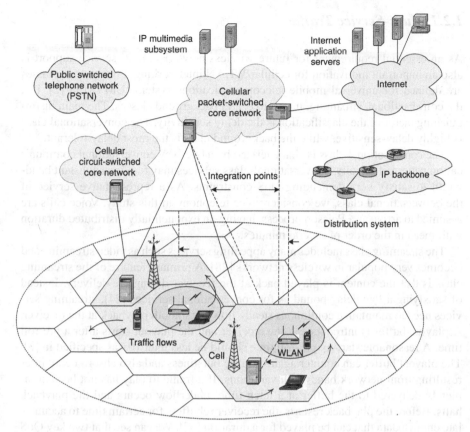

Fig. 1.1 System model for a cellular/WLAN integration network

to 14 Mbit/s. However, these broadband wireless technologies are still not widely applied to the cellular networks in operation. Also, the deployment of microcells or picocells in hotspots is not so cost-effective as WLAN deployment. Hence, cellular networks usually have a much smaller cell capacity than WLANs.

As shown in Fig. 1.1, the cellular network provides ubiquitous connectivity over wide-area coverage, while WLANs are deployed disjointly in hotspot local areas. Assume that there is one overlay WLAN in a cell and the WLAN is overlaid with one and only one cell. The cell and its overlay WLAN are referred to as a *cell/WLAN cluster*. The mobile devices are dual-mode and equipped with network interfaces to both the cellular network and WLAN [39]. Thus, both cellular access and WLAN access are available to dual-mode mobiles within the WLAN-covered areas, which are referred to as *double-coverage areas*. Since the two networks operate at different frequency bands, the two network interfaces can be active simultaneously to assist vertical handoff [47]. In contrast, the areas with only cellular coverage are referred to as *cellular-only areas*.

1.2.1 Multi-Service Traffic

As an essential requirement for future wireless networks, multi-service support is also an important motivation for cellular/WLAN interworking. Four service classes are defined for universal mobile telecommunications system (UMTS) in [2], i.e., the conversational, streaming, interactive, and background classes. The main distinguishing factor of the classification is the delay sensitivity. The conversational class is highly delay-sensitive, while the background class is the most delay tolerant.

The conversational class is characterized by a two-way conversational communication pattern. Typically, conversational-class services may require a constant bandwidth to satisfy very demanding QoS constraints. As a representative service of the conversational class, we consider voice telephony in this study. Voice calls are assumed to arrive as a Poisson process, having an exponentially distributed duration with mean in the order of several minutes.

The streaming class includes many appealing services such as video streaming and becomes very popular in wireless networks [28]. A primary feature of the streaming class is that the content is played back at the receiver during the delivery. Instead of satisfying a low delay bound as for conversational services [24], streaming services need to maintain a continuous steady flow for smooth playback at the receiver. A playout buffer is introduced at the receiver and the playout starts after a pre-roll time. A reasonable start-up pre-roll time should be less than 10 s as specified in [3]. The playout buffer can counter against traffic burstiness and also absorb delay jitter resulting from network bandwidth variations. If a frame to play has not been completely delivered to the buffer at a fetch time, underflow occurs and the playback halts. Before the playback restarts, the receiver rebuffers for certain time to accumulate enough data that can be played for a duration [37]. We can see that two key QoS metrics are the pre-roll time and rebuffer time.

The interactive class comprises non-real time services with a request-response pattern, such as Web browsing, voice messaging, and file transfer. Invoked independently by a large number of independent users, call arrivals are assumed to be Poisson. As non-real time services, interactive data calls are tolerant of elastic bandwidth. The transfer delay (also known as response time) is a main QoS criterion to measure the responsiveness, e.g., how fast a Web page is successfully downloaded and appears after it has been requested, or equivalently the call throughput, which is the ratio of file size over the transfer delay [36]. Although the transfer delay should be bounded to maintain fluent interactions, the delay requirement is much less stringent than that of conversational services [3]. A transfer delay of 2–4 s per page is acceptable for Web browsing and a desirable target is 0.5 s.

Interactive service exhibits the on-off dynamics [1] shown in Fig. 1.2. If the download of a Web page or a data file is viewed as a packet data call, an interactive session consists of a sequence of data calls (the "on" phases). After downloading a Web page, the user may take a "reading time" (the "off" phase) before requesting the next page, and finish the session after reading a number of pages. The reading time is

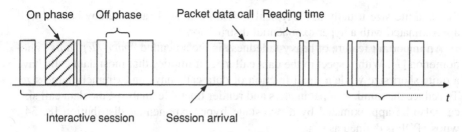

Fig. 1.2 Structure of interactive data sessions

assumed to be exponentially distributed, and the number of data calls in a session is geometrically distributed.

For interactive data services such as Web browsing and file transfer, it is observed that the packet-level traffic presents asymptotic self-similarity and high variability over a wide range of time scales [9]. This property is attributed to the heavy-tailed file size and burstiness induced by traffic control mechanisms such as closed-loop congestion control [13]. Although the complex packet-level traffic characterization may complicate performance analysis, the QoS metrics of interest are actually more dependent on higher flow-level or session-level behaviors and less relevant to packet-level dynamics. For example, the mean response time depends on flow fluctuation and bandwidth sharing manner among in-progress flows [36].

As an essential call-level traffic characteristic, the heavy-tailedness of data file size has been extensively studied in the literature. The statistics of real file size can be captured with heavy-tailed distributions. In [35], the data file size L_d is modeled by a Weibull distribution, whose probability density function (PDF) is given by

$$f_{L_d}(x) = \frac{\alpha_d}{\beta_d} \left(\frac{x}{\beta_d} \right)^{\alpha_d - 1} e^{-(x/\beta_d)^{\alpha_d}} \triangleq W_b(x, \alpha_d, \beta_d)$$
$$0 < \alpha_d \leq 1, \quad \beta_d > 0, \quad x > 0 \tag{1.1}$$

where α_d is the shape parameter and β_d is the scale parameter. The PDF of the Weibull distribution is denoted by $W_b(x, \alpha_d, \beta_d)$ for simplicity. The mean of L_d is given by $E[L_d] = \beta_d \, \Gamma(1 + \frac{1}{\alpha_d}) \triangleq f_d$, where $\Gamma(\cdot)$ is the Gamma function. The exponential distribution is actually a special case of the Weibull distribution with $\alpha_d = 1$, while the Weibull distribution is heavy-tailed if $0 < \alpha_d < 1$. The smaller the Weibull factor α_d, the heavier the tail that occurs in a given Weibull distribution.

To render effective and tractable analysis, it is proposed in [12] to fit a large class of heavy-tailed distributions with hyper-exponential distributions. Hyper-exponential distributions are a special class of phase-type distributions, which are a very general mixture of exponential distributions and have been used to approximate general distributions. In particular, for a distribution with a coefficient of variance (CV) larger than 1, a hyper-exponential distribution can be used since the CV of a hyper-exponential distribution is always larger than 1. As observed from real measurements, the data file size usually has a typical CV larger than 1, and thereby can be well approximated with a hyper-exponential distribution.

An important feature of heavy-tailedness is the so-called *"mice-elephants"* phenomenon [7]. With respect to the data call size, it implies that most data calls have a quite short size while a small fraction of data calls have an extremely large size. To reduce the number of parameters and render tractable analysis, the data call size can also be approximated by a two-stage hyper-exponential distribution [8, 54], whose PDF is defined as

$$f_{L_d}(x) = \frac{b}{b+1} \cdot \frac{b}{f_d} \exp\left(-\frac{b}{f_d}x\right) + \frac{1}{b+1} \cdot \frac{1}{bf_d} \exp\left(-\frac{1}{bf_d}x\right), \quad b \geq 1, x > 0 \tag{1.2}$$

where the parameters b and f_d can be obtained by fitting the first and second moments. In particular, b can completely characterize the *"mice-elephants"* feature. A larger value of b corresponds to a data call size with a higher variability. Furthermore, since the hyper-exponential distribution consists of a linear mixture of exponentials, the analytical study involving (1.2) can be extended to higher-order hyper-exponential distributions with more exponential components, which can more accurately approach the original heavy-tailed distribution. Hence, hyper-exponential approximation can not only provide analytical tractability but also well capture the essential properties of heavy-tailed distributions.

1.2.2 Location-Dependent User Mobility

It is known that most WLANs are deployed in indoor environments like cafés, offices, and airports. Users within these areas are mostly static or only maintain a pedestrian-level mobility. Thus, it becomes not reasonable to apply a homogeneous mobility model for mobiles within the coverage of a large cell. For example, when a user drives to the office, its mobility level may change from a high vehicular speed on the highway to being almost static in the office. To statistically characterize user mobility, user residence time should vary with the location, which is either within the cellular-only coverage or double coverage.

With a statistical equilibrium assumption, we focus on a single cell with an overlay WLAN, i.e., a cell/WLAN cluster. As shown in [53], the indoor deployment and low user mobility result in a heavy-tailed user residence time within a WLAN. To avoid the complexity of directly applying heavy-tailed distributions in performance analysis, the user residence time within a WLAN, denoted by T_r^w, is modeled with an approximate hyper-exponential distribution [54], whose PDF is given by

$$f_{T_r^w}(t) = \frac{a}{a+1} \cdot a\eta^w \exp\left(-a\eta^w t\right) + \frac{1}{a+1} \cdot \frac{\eta^w}{a} \exp\left(-\frac{\eta^w}{a}t\right), \quad a \geq 1, t > 0 \tag{1.3}$$

where the mean and squared coefficient of variance are respectively

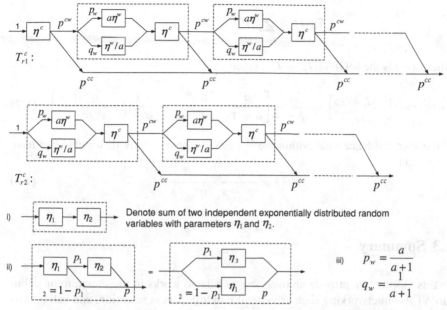

i) Denote sum of two independent exponentially distributed random variables with parameters η_1 and η_2.

ii) ... iii) $p_w = \dfrac{a}{a+1}$, $q_w = \dfrac{1}{a+1}$

Denote a random variable which with probability p_2 follows an exponential distribution with parameter η_1 and with probability p_1 follows a generalized hyperexponential distribution with parameter η_3 (the sum of two exponential random variables with parameters η_1 and η_2).

Fig. 1.3 Modeling of user mobility within a cell/WLAN cluster

$$E[T_r^w] = (\eta^w)^{-1}, \quad \frac{\mathrm{Var}[(T_r^w)]}{E^2[T_r^w]} = 2a + \frac{2}{a} - 3 \triangleq C_{v,T_r^w}^2. \tag{1.4}$$

This model well captures the "*mice-elephants*" property of heavy-tailedness. A large fraction $\frac{a}{a+1}$ of the users stay within the WLAN for a mean time $(a\eta^w)^{-1}$, while the other $\frac{1}{a+1}$ of the users have a mean residence time of (a/η^w). Increasing the parameter a results in T_r^w with higher variability.

On the other hand, the user residence time in the area of a cell with only cellular access, denoted by T_r^c, is assumed to be exponentially distributed with parameter η^c. Users moving out of the cellular-only area enter neighboring cells with a probability p^{cc} and enter the coverage of the overlay WLAN in the target cell with a probability $p^{cw} = 1 - p^{cc}$. Therefore, the residence time of users admitted in the cell follows more complicate phase-type distributions as shown in Fig. 1.3. Let T_{r1}^c and T_{r2}^c denote the cell residence time of a call from the cellular-only area and that of a call from the double-coverage area, respectively. The moment generating functions (MGFs) of T_{r1}^c and T_{r2}^c are derived from Fig. 1.3 as

$$\Phi_1(s) = \sum_{i=1}^{\infty} (p^{cw})^{i-1} p^{cc} \frac{\eta^c}{\eta^c - s} [\psi(s)]^{i-1} \tag{1.5}$$

$$\Phi_2(s) = \sum_{i=1}^{\infty} (p^{cw})^{i-1} p^{cc} [\psi(s)]^i \qquad (1.6)$$

where $\psi(\cdot)$ is the MGF of $T_r^c + T_r^w$, given by

$$\psi(s) = \mathrm{E}\left[e^{s(T_r^c + T_r^w)}\right] = \frac{\eta^c}{\eta^c - s} \left[\frac{a}{a+1} \cdot \frac{a\eta^w}{a\eta^w - s} + \frac{1}{a+1} \cdot \frac{\eta^w/a}{\eta^w/a - s}\right]. \qquad (1.7)$$

If the user residence time within the WLAN is exponentially distributed, we have $a = 1$ and

$$\psi(s) = \frac{\eta^c}{\eta^c - s} \cdot \frac{\eta^w}{\eta^w - s}. \qquad (1.8)$$

1.3 Summary

In this chapter, we provide an overview of related works in the literature on cellular/WLAN interworking such as vertical handoff, access selection, and call admission control. Many studies in these areas do not address multi-service support or neglect the unique characteristics of the integrated network. As the cellular network and WLAN differ in capacity, mobility support, and QoS provisioning, an appropriate system model is important to obtain accurate observations on cellular/WLAN interworking. In the following chapters, we present three state-of-the-art solutions to cellular/WLAN interworking. The three representative interworking schemes explore the fundamental issues such as resource allocation, access selection, and call admission control in a comprehensive fashion. A practical system model is taken into account so as to capture multi-service traffic characteristics and location-dependent user mobility. We start with a straightforward research allocation scheme, in Chap. 2, for the cellular/WLAN integrated network to support integrated voice and data services.

Chapter 2
Resource Allocation for Integrated Voice and Data Services

In the interworking between a cellular network and WLANs, there is a two-tier overlaying structure which offers both cellular and WLAN access to a dual-mode mobile station (MS) within the WLAN-covered area. Ubiquitous coverage is provided by the cellular network (higher-tier), while WLANs (lower-tier) are deployed in disjoint hot-spot areas. Then, there comes the access selection problem of how to properly admit incoming traffic to the cell or WLAN. Specially, a preferred target network, either a cellular cell or a WLAN, should be first selected based on various decision criteria taking into account factors such as service type and network conditions of the two networks. A service request rejected by its first-choice network can just leave the system or further try to access the other network [26]. Due to user mobility and access selection in the overlaying area, the underlying network serving a user may alternate dynamically between the cellular network and WLANs. Due to heterogeneous underlying technologies, the admission choice can have a significant impact on overall resource utilization and QoS satisfaction. In this chapter, we introduce an easy-to-implement resource allocation scheme for voice and data services over the cellular/WLAN integrated network.

2.1 Bandwidth Sharing for Voice and Data Services

Consider real-time voice telephony and interactive data services (such as Web browsing). Each voice call requires a constant bandwidth to meet its strict delay requirement, while data service is adaptive to elastic bandwidth. Each voice call has two voice flows from and to the MS, while each data call has a one-way data flow to the MS. Moreover, we assume the mean arrival rate of new voice (data) calls in the cellular-only area λ_{v1} (λ_{d1}) is proportional to that in the double-coverage area λ_{v2} (λ_{d2}). As WLANs are usually deployed in hot-spot areas, on average, the traffic density in the double-coverage area is higher than that in the cellular-only area. In addition to new traffic, there are also horizontal handoffs between neighboring cells and vertical handoffs between a WLAN and its overlaying cell.

W. Song and W. Zhuang, *Interworking of Wireless LANs and Cellular Networks*,
SpringerBriefs in Computer Science, DOI: 10.1007/978-1-4614-4379-7_2,
© The Author(s) 2012

In the cellular network, with the aid of base stations, the restricted access policy [29] can be applied. With this policy, voice is only allowed to occupy certain bandwidth, while the remaining bandwidth is dedicated to data. All the bandwidth unused by current voice traffic is shared equally by existing data calls. That is, a processor sharing (PS) service discipline is applied to data traffic, and the total bandwidth occupied by data traffic dynamically varies with voice traffic. This policy is shown to achieve higher utilization than complete sharing and complete partitioning [33] and to offer each service certain QoS protection against the other. In WLANs, with contention-based random access, multiple services are supported in complete sharing. Admission control is necessary to limit both voice and data users in service. Otherwise, the intra-service interference from users of the same service type or inter-service interference from users of the other service type may severely degrade the system performance.

2.1.1 Allocation of WLAN and Cell Bandwidth

To apply joint resource allocation for the integrated network, we need to first analyze the capacity of each network for voice and data services. With centralized control and bandwidth reservation, the cell capacity is relatively easy to analyze, while the contention-based access and complete resource sharing in WLANs complicate the WLAN capacity analysis.

Suppose there are n_v^w voice calls and n_d^w data calls admitted in a WLAN. Packets from a voice flow are assumed to arrive with a constant rate, λ_v^p. For Web browsing, the data file to be transmitted is usually pre-stored in a server. Therefore, it is reasonable to consider that there is always traffic during the lifetime of a data call. Data transmission follows the optional request-to-send and clear-to-send (RTS-CTS) handshaking for channel access, while voice flows use the basic carrier sensing multiple access with collision avoidance (CSMA/CA) mechanism due to the small payload size of voice packets. Following the approach in [43], we can derive the service rates for packets from one voice and data flow, denoted by $\xi_v^w(n_v^w, n_d^w)$ and $\xi_d^w(n_v^w, n_d^w)$, respectively. To satisfy the real-time requirement of voice traffic, the service rate of a voice flow needs to be greater than the voice packet arrival rate. Thus, the following constraint should be met: $\xi_v^w(n_v^w, n_d^w) > (1 + \delta)\lambda_v^p$, where δ is a design parameter to be determined experimentally. Given this stability constraint, we can get the capacity region, i.e., the feasible set of (n_v^w, n_d^w) vectors, and the corresponding data service rate $\xi_d^w(n_v^w, n_d^w)$ for each (n_v^w, n_d^w) vector in the capacity region.

Given the cell bandwidth C^c and total voice traffic load, the minimum bandwidth needed to meet the requirements of voice call blocking and dropping probabilities can be obtained as $R_v^c \cdot N_v^c$, where R_v^c is the bandwidth requirement of a voice call and N_v^c ($\leq \lfloor \frac{C^c}{R_v^c} \rfloor$) is the maximum number of voice calls allowed in a cell. Moreover, because only cellular access is available in the cellular-only area, randomized guard

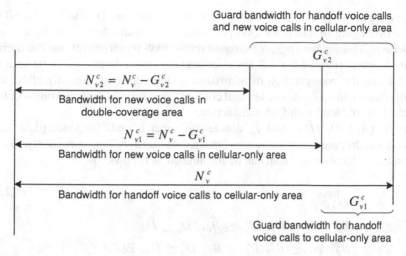

Fig. 2.1 Randomized guard channel policy for voice in the cell

channel policy is applied to give the new and handoff traffic in this area a priority to access the cell bandwidth over the traffic in the double-coverage area. Because the call blocking and dropping probabilities are very sensitive to the amount of reserved bandwidth, the guard bandwidth for high-priority voice traffic is randomized instead of an integer number of guard channels. As shown in Fig. 2.1, the voice admission region of the cell is given by $(N_v^c, G_{v1}^c, G_{v2}^c)$, in which G_{v2}^c $(\leq N_v^c)$ is a real number representing a randomized number of guard channels (guard bandwidth) dedicated to new and handoff voice traffic in the cellular-only area and G_{v1}^c $(\leq G_{v2}^c)$ is the guard bandwidth reserved only for handoff voice traffic in this area. On the other hand, the remaining cell bandwidth $(C^c - R_v^c \cdot N_v^c)$ is dedicated to data. All on-going data calls equally share the bandwidth unused by voice, and the data service rate is dynamically adjusted with call arrivals and departures. In addition, data traffic is prioritized similarly to voice based on user location area and new/handoff call differentiation. The data admission region of the cell is given by $(N_d^c, G_{d1}^c, G_{d2}^c)$.

2.1.2 Formulation of Resource Allocation Problem

The WLAN capacity region is derived in terms of (n_v^w, n_d^w) vectors. It is found that the data service rate $\xi_d^w(n_v^w, n_d^w)$ decreases dramatically with more voice calls (i.e., a larger n_v^w), which implies the inefficient voice support of WLANs. To prevent the WLAN from operating in that inefficiency region, instead of directly applying the two-dimensional capacity region as admission criteria, we limit the maximum number of voice calls and that of data calls admitted in the WLAN by N_v^w and N_d^w, respectively. The WLAN admission region (N_v^w, N_d^w) is chosen within the WLAN capacity region to guarantee packet-level QoS satisfaction. Due to user mobility and the overlaying structure, the QoS performance is jointly determined by the cell and WLAN. Thus, given (N_v^w, N_d^w), based on the QoS requirements, we can derive the admission regions of the cell for voice and data accordingly. It can be seen in Sect. 2.4 that the configuration of admission regions does significantly affect the overall resource utilization. The optimal configuration assures a maximization of the acceptable traffic load with QoS satisfaction.

Let $\tilde{B}_v$ ($\tilde{B}_d$), $\tilde{D}_v$ ($\tilde{D}_d$), and $\tilde{T}_d$ denote the upper bounds for voice (data) call blocking and dropping probabilities and mean data transfer time, respectively. Then, the resource allocation problem can be formulated as follows:

$$\max_{(N_v^w, N_d^w)} \lambda_d, \ s.t. \tag{2.1}$$

$$B_v^w \cdot B_{v2}^c \leq \tilde{B}_v, \ \ B_{v1}^c \leq \tilde{B}_v, \ \ D_v^c \leq \tilde{D}_v$$

$$B_d^w \cdot B_{d2}^c \leq \tilde{B}_d, \ \ B_{d1}^c \leq \tilde{B}_d, \ \ D_d^c \leq \tilde{D}_d, \ \mathrm{E}[T_d] \leq \tilde{T}_d$$

where λ_d ($= \lambda_{d1} + \lambda_{d2}$) is the mean data call arrival rate in the cell cluster-covered area, B_{v1}^c and B_{v2}^c (B_{d1}^c and B_{d2}^c) are the blocking probabilities of the cell for new voice (data) calls in the cellular-only area and double-coverage area, respectively, D_v^c is the voice handoff dropping probability of the cell, B_v^w (B_d^w) is the probability that a voice (data) call is blocked in the WLAN, and $\mathrm{E}[T_d]$ is the mean data transfer time. Since we fix the voice call arrival rates for simplicity, the maximization of λ_d implies a maximization of the total acceptable traffic load and resource utilization.

2.2 Voice Performance of WLAN-First Scheme

In this section, we investigate a simple and easy-to-implement access selection strategy, referred to as *WLAN-first scheme*, where WLANs are always preferred by all services whenever the WLAN access is available, so as to take advantage of the low cost and large bandwidth of WLANs. An incoming service request rejected by a WLAN overflows to the cellular network to request admission if it is a new call, or remains in the cellular network if it is an ongoing call carried by the overlaying cell. Although the WLAN-first scheme is a straightforward approach, an in-depth analysis is very meaningful to examine how various services affect the resource allocation and QoS support in a cellular/WLAN integrated network.

Because a voice call duration is of the order of minutes, while a data call is required to finish transmission within seconds, the number of voice calls fluctuates much more slowly than that of data calls. No voice call arrival or departure is assumed during a data call duration. In particular, this limiting behavior for a Markov chain is referred to as *nearly complete decomposability* [33]. Let $(k_v^w, k_{v1}^c, k_{v2}^c)$ denote the state of voice traffic in a cell cluster, where k_v^w, k_{v1}^c, and k_{v2}^c are the numbers of voice calls admitted to the WLAN, to the cell from the cellular-only area, and to the cell from the double-coverage area, respectively. The number of voice calls in the WLAN can be described by a birth-death process with respect to k_v^w. In this study, we first consider a simple case that the user residence time T_r^w in the double-coverage area is exponentially distributed, i.e., $a = 1$ in (1.3). Since both voice call duration T_v and user residence time T_r^w are exponentially distributed, the voice channel occupancy time $\min(T_v, T_r^w)$ is exponential with mean $1/(\mu_v + \eta^w)$, where $1/\mu_v$ is the mean voice call duration. Then, the steady-state probability of k voice calls in the WLAN is obtained based on an $M/M/K/K$ loss system, given by

$$\pi_v^w(k) = \frac{[(\lambda_{v2} + \lambda_{hv}^{cw})/(\mu_v + \eta^w)]^k/k!}{\sum_{i=0}^{N_v^w} [(\lambda_{v2} + \lambda_{hv}^{cw})/(\mu_v + \eta^w)]^i/i!}, \quad 0 \le k \le N_v^w \tag{2.2}$$

where λ_{v2} and λ_{hv}^{cw} are the mean arrival rates of new and handoff voice calls to the WLAN, respectively. Thus, the voice call blocking probability in the WLAN is $B_v^w = \pi_v^w(N_v^w)$.

Next, we analyze the voice performance in a cell, which is more complex due to traffic prioritization. We draw in Fig. 2.2 the state transition diagram of (k_{v1}^c, k_{v2}^c), which is divided into several areas for illustration purpose and in each area only one example transition is shown with respect to k_{v1}^c and k_{v2}^c, respectively. The state-dependent transition rates are conditioned on k_v^w and derived as follows.

In general, suppose X and Y are two independent random variables with $X \sim \exp(\lambda)$. Then,

$$P[X > Y] = \int_0^\infty f_Y(y)dy \int_y^\infty \lambda e^{-\lambda x}dx = \int_0^\infty f_Y(y)e^{-\lambda y}dy = \Psi_Y(-\lambda) \tag{2.3}$$

where $f_Y(\cdot)$, $F_Y(\cdot)$, and $\Psi_Y(\cdot)$ are the PDF, cumulative probability function (CDF), and MGF of Y, respectively. Letting $Z = \min(X, Y)$, the PDF of Z is given by

$$\begin{aligned} f_Z(z) &= f_X(z)[1 - F_Y(z)] + f_Y(z)[1 - F_X(z)] \\ &= f_X(z) + f_Y(z) - [f_Y(z)F_X(z) + f_X(z)F_Y(z)] \end{aligned} \tag{2.4}$$

where $f_X(\cdot)$ and $F_X(\cdot)$ denote the PDF and CDF of X, respectively. Then, the mean value of Z is

$$E[Z] = \int_0^\infty z f_Z(z)dz = E[X] + E[Y] - \int z[F_X(z)F_Y(z)]'dz$$

$$= E[X] - \int_0^\infty f_Y(y)\frac{1}{\lambda}e^{-\lambda y}dy = \frac{1}{\lambda} - \frac{1}{\lambda}\Psi_Y(-\lambda) = \left[\frac{\lambda}{1 - \Psi_Y(-\lambda)}\right]^{-1}.$$

$$\tag{2.5}$$

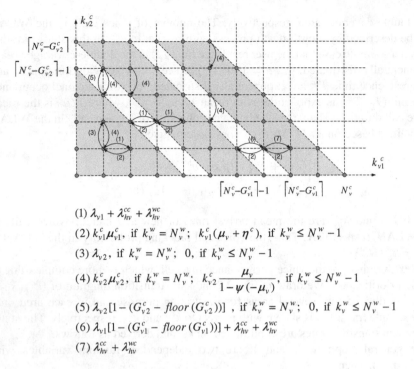

(1) $\lambda_{v1} + \lambda_{hv}^{cc} + \lambda_{hv}^{wc}$

(2) $k_{v1}^c \mu_{v1}^c$, if $k_v^w = N_v^w$; $k_{v1}^c(\mu_v + \eta^c)$, if $k_v^w \le N_v^w - 1$

(3) λ_{v2}, if $k_v^w = N_v^w$; 0, if $k_v^w \le N_v^w - 1$

(4) $k_{v2}^c \mu_{v2}^c$, if $k_v^w = N_v^w$; $k_{v2}^c \dfrac{\mu_v}{1 - \psi(-\mu_v)}$, if $k_v^w \le N_v^w - 1$

(5) $\lambda_{v2}[1 - (G_{v2}^c - floor\,(G_{v2}^c))]$, if $k_v^w = N_v^w$; 0, if $k_v^w \le N_v^w - 1$

(6) $\lambda_{v1}[1 - (G_{v1}^c - floor\,(G_{v1}^c))] + \lambda_{hv}^{cc} + \lambda_{hv}^{wc}$

(7) $\lambda_{hv}^{cc} + \lambda_{hv}^{wc}$

Fig. 2.2 State transition diagram for voice in the cell

For new and handoff voice calls in the cellular-only area, the channel occupancy time is $\min(T_v, T_{r1}^c)$. Based on (1.5) and (2.5), its mean value can be derived by

$$\mathrm{E}[\min(T_v, T_{r1}^c)] = \frac{1}{\mu_v} - \frac{1}{\mu_v} \sum_{i=1}^{\infty} (p^{cw})^{i-1} p^{cc} \frac{\eta^c}{\eta^c + \mu_v} [\psi(-\mu_v)]^{i-1}$$

$$= \frac{1}{\mu_v} - \frac{1}{\mu_v} p^{cc} \frac{\eta^c}{\eta^c + \mu_v} \frac{1}{1 - p^{cw}\psi(-\mu_v)} \triangleq \frac{1}{\mu_{v1}^c} \qquad (2.6)$$

where

$$\psi(-\mu_v) = \frac{\eta^c}{\eta^c + \mu_v} \cdot \frac{\eta^w}{\eta^w + \mu_v}.$$

Similarly, the channel occupancy time for new voice calls in the double-coverage area is $\min(T_v, T_{r2}^c)$ with mean value

$$\mathrm{E}[\min(T_v, T_{r2}^c)] = \frac{1}{\mu_v} - \frac{1}{\mu_v} \sum_{i=1}^{\infty} (p^{cw})^{i-1} p^{cc} [\psi(-\mu_v)]^i$$

$$= \frac{1}{\mu_v} - \frac{1}{\mu_v} p^{cc} \frac{\psi(-\mu_v)}{1 - p^{cw}\psi(-\mu_v)} \triangleq \frac{1}{\mu_{v2}^c}. \tag{2.7}$$

To further simplify analysis, we average the transition rates of (k_{v1}^c, k_{v2}^c) over k_v^w depending on whether there is enough free capacity in the WLAN for an arriving voice call. Then, the departure rate from state (k_{v1}^c, k_{v2}^c) to state $(k_{v1}^c - 1, k_{v2}^c)$ $(k_{v1}^c \geq 1)$ and the departure rate from state (k_{v1}^c, k_{v2}^c) to state $(k_{v1}^c, k_{v2}^c - 1)$ $(k_{v2}^c \geq 1)$ are respectively approximated by

$$\tilde{\mu}_{v1}^c = B_v^w k_{v1}^c \mu_{v1}^c + (1 - B_v^w) k_{v1}^c (\mu_v + \eta^c) \tag{2.8}$$

$$\tilde{\mu}_{v2}^c = B_v^w k_{v2}^c \mu_{v2}^c + (1 - B_v^w) k_{v2}^c \frac{\mu_v}{1 - \psi(-\mu_v)}. \tag{2.9}$$

As indicated by $\tilde{\mu}_{v1}^c$ and $\tilde{\mu}_{v2}^c$, voice traffic admitted to the cell from the cellular-only area and from the double-coverage area has different mean channel occupancy times approximated by $(\tilde{\mu}_{v1}^c)^{-1}$ and $(\tilde{\mu}_{v2}^c)^{-1}$, respectively. Hence, the cell can be viewed as a multi-service loss system [38]. A product-form state distribution exists and is insensitive to service time distributions, provided that the resource sharing among services is under coordinate convex policies. This requires that transitions between states come in pairs. For loss systems with trunk reservation (e.g., the guard channel policy), the insensitivity property and product-form solution are destroyed due to the one-way transitions at some states. In [55], the state distribution is approximated with a recursive method, which is shown to be accurate for a wide range of traffic intensities and when the service rates (such as $\tilde{\mu}_{v1}^c$ and $\tilde{\mu}_{v2}^c$) do not greatly differ from each other. Moreover, the blocking probabilities are *almost* insensitive to service time distributions. Hence, we use the recursive approximation in [55] to obtain $\pi_v^c(k)$ (i.e., the steady-state probability of k voice calls admitted into the cell) as follows[1]:

$$\pi_v^c(k) = \left(\frac{\lambda_{v1}^c}{\tilde{\mu}_{v1}^c} + \frac{\lambda_{nv2}^c}{\tilde{\mu}_{v2}^c}\right)^k \frac{\pi_v^c(0)}{k!}, \quad 0 \leq k \leq \lfloor N_{v2}^c \rfloor$$

$$\pi_v^c(k) = \left(\frac{\lambda_{v1}^c}{\tilde{\mu}_{v1}^c} + \frac{\lambda_{nv2}^c}{\tilde{\mu}_{v2}^c}\right)^{\lfloor N_{v2}^c \rfloor} \rho_{v2} \left(\frac{\lambda_{v1}^c}{\tilde{\mu}_{v1}^c}\right)^{k - \lfloor N_{v2}^c \rfloor - 1} \frac{\pi_v^c(0)}{k!}, \quad \lfloor N_{v2}^c \rfloor + 1 \leq k \leq \lfloor N_{v1}^c \rfloor$$

$$\pi_v^c(k) = \left(\frac{\lambda_{v1}^c}{\tilde{\mu}_{v1}^c} + \frac{\lambda_{nv2}^c}{\tilde{\mu}_{v2}^c}\right)^{\lfloor N_{v2}^c \rfloor} \rho_{v2} \left(\frac{\lambda_{v1}^c}{\tilde{\mu}_{v1}^c}\right)^{\lfloor N_{v1}^c \rfloor - \lfloor N_{v2}^c \rfloor - 1} \rho_{v1} \left(\frac{\lambda_{hv}^c}{\tilde{\mu}_{v1}^c}\right)^{k - \lfloor N_{v1}^c \rfloor - 1} \frac{\pi_v^c(0)}{k!},$$

$$\lfloor N_{v1}^c \rfloor + 1 \leq k \leq N_v^c \tag{2.10}$$

---

[1] The expression is given under the condition that $\lfloor G_{v1}^c \rfloor \leq \lfloor G_{v2}^c \rfloor - 1$ and $\lfloor G_{v1}^c \rfloor \geq 1$. When $\lfloor G_{v1}^c \rfloor = 0$ or $\lfloor G_{v1}^c \rfloor = \lfloor G_{v2}^c \rfloor$, the expression can be adjusted accordingly based on the recursive method in [55].

where $\pi_v^c(0) = 1/C_1$ with C_1 being a normalization constant,[2] and

$$N_{v1}^c = N_v^c - G_{v1}^c, \quad N_{v2}^c = N_v^c - G_{v2}^c$$

$$\lambda_{v1}^c = \lambda_{v1} + \lambda_{hv}^{wc} + \lambda_{hv}^{cc}, \quad \lambda_{hv}^c = \lambda_{hv}^{wc} + \lambda_{hv}^{cc}, \quad \lambda_{nv2}^c = B_v^w \lambda_{v2}$$

$$\rho_{v1} = \frac{[1 - (G_{v1}^c - \lfloor G_{v1}^c \rfloor)]\lambda_{v1} + \lambda_{hv}^c}{\tilde{\mu}_{v1}^c},$$

$$\rho_{v2} = \frac{\lambda_{v1}^c}{\tilde{\mu}_{v1}^c} + \frac{[1 - (G_{v2}^c - \lfloor G_{v2}^c \rfloor)]\lambda_{nv2}^c}{\tilde{\mu}_{v2}^c}.$$

Thus, the voice call blocking and dropping probabilities of the cell are given by

$$B_{v2}^c = (G_{v2}^c - \lfloor G_{v2}^c \rfloor) \, \pi_v^c(\lfloor N_{v2}^c \rfloor) + \sum_{i=\lfloor N_{v2}^c \rfloor + 1}^{N_v^c} \pi_v^c(i) \qquad (2.11)$$

$$B_{v1}^c = (G_{v1}^c - \lfloor G_{v1}^c \rfloor) \, \pi_v^c(\lfloor N_{v1}^c \rfloor) + \sum_{i=\lfloor N_{v1}^c \rfloor + 1}^{N_v^c} \pi_v^c(i) \qquad (2.12)$$

$$D_v^c = \pi_v^c(N_v^c). \qquad (2.13)$$

2.3 Data Performance of WLAN-First Scheme

Under the limiting case that the time scale of voice calls is much larger than that of data calls, the analysis for data traffic can be approximately decoupled from that of voice [33].

2.3.1 Mean Data Transfer Time

First, we consider the performance of data service in the cell. Since all the bandwidth unused by voice traffic is shared equally by existing data calls, a cell behaves like an $M/G/1/K - PS$ queue, whose service capacity is $(C^c - iR_v^c)$ with a probability $\pi_v^c(i), i = 0, 1, \ldots, N_v^c$. Thus, the expected duration of a data call carried by the cell is approximated by [10]

$$\mathrm{E}[T_d^c] = \sum_{i=0}^{N_v^c} \pi_v^c(i) \frac{\rho_d^c(i)^{N_d^c+1}\left[N_d^c\rho_d^c(i) - N_d^c - 1\right] + \rho_d^c(i)}{\lambda_d^c\left[1 - \rho_d^c(i)^{N_d^c}\right]\left[1 - \rho_d^c(i)\right]} \qquad (2.14)$$

---

[2] C_2, C_3, and C_4 used in the following are all normalization constants.

where

$$\lambda_d^c = \lambda_{d1} + \lambda_{nd2}^c + \lambda_{hd}^{wc} + \lambda_{hd}^{cc}, \quad \lambda_{nd2}^c = B_d^w \lambda_{d2}, \quad \rho_d^c(i) = \frac{\lambda_d^c f_d}{C^c - i R_v^c}$$

and λ_{nd2}^c is the mean arrival rate of new data calls overflowed to the cell in the double-coverage area, f_d is the mean data file size. Given in (2.14) is actually an upper bound for the mean transfer time of a data call with exponentially distributed size [10]. When data traffic evolves rapidly with respect to voice traffic, i.e., the number of data calls can attain its stationary regime given by an $M/G/1/K - PS$ queue with service capacity $(C^c - i R_v^c)$, the upper bound can be used to approximate the mean data transfer time.

Because a data call may be carried by different cells and/or WLANs during its lifetime, its overall performance depends on both networks. Next, we analyze the expected data call duration when a data call is carried by a WLAN. In the WLAN, the data service rate is state-dependent due to the complete resource sharing between voice and data traffic. The probability of j data calls carried by the WLAN is given by

$$\tilde{\pi}_d^w(j) = \sum_{i=0}^{N_v^w} \left[\pi_v^w(i) \tilde{\pi}_d^w(0) \frac{(\lambda_d^w)^j}{\prod_{l=1}^{j} l \chi_d^w(i, l)} \right], \quad j = 1, 2, \ldots, N_d^w \quad (2.15)$$

$$\tilde{\pi}_d^w(0) = 1/C_2, \quad \lambda_d^w = \lambda_{d2} + \lambda_{hd}^{cw}, \quad \chi_d^w(i, l) = \frac{\xi_d^w(i, l)}{f_d}$$

where $\pi_v^w(i)$ is given by (2.2), $\chi_d^w(i, l)$ is the service rate for one data call with i voice calls and l data calls in the WLAN, λ_{d2} and λ_{hd}^{cw} are the mean arrival rates of new and handoff data calls to the WLAN, respectively. Using the Little's law, the expected duration of a data call carried by the WLAN can be obtained as

$$E[T_d^w] = \frac{1}{\lambda_d^w(1 - B_d^w)} \sum_{j=0}^{N_d^w} j \tilde{\pi}_d^w(j) \quad (2.16)$$

where B_d^w is the data call blocking probability of the WLAN.

2.3.2 Blocking and Dropping Probabilities of Data Calls

Consider the state that there are i voice calls and j data calls in a cell. For data calls admitted to the cell from the cellular-only area, by averaging over the WLAN state, we approximate the departure rate from state (i, j) to state $(i, j - 1)$ $(j \geq 1)$ by

$$\tilde{\mu}_{d1}^{c}(i, j) = B_d^w j \mu_{d1}^c(i, j) + (1 - B_d^w) j [\nu_d^c(i, j) + \eta^c] \tag{2.17}$$

where

$$\nu_d^c(i, j) = \frac{C^c - i R_v^c}{j f_d} \tag{2.18}$$

and $\mu_{d1}^c(i, j)$ is the inverse of mean cell bandwidth occupancy time of data calls when there is not enough free capacity in the WLAN, which can be obtained from (2.5) as

$$\mu_{d1}^c(i, j) = \frac{j \nu_d^c(i, j)}{1 - \Phi_1(-\nu_d^c(i, j))}. \tag{2.19}$$

Similarly, for data calls admitted to the cell from the double-coverage area, the departure rate from state (i, j) to state $(i, j - 1)$ $(j \geq 1)$ is

$$\tilde{\mu}_{d2}^c(i, j) = B_d^w \cdot j \mu_{d2}^c(i, j) + (1 - B_d^w) \cdot \frac{j \nu_d^c(i, j)}{1 - \psi(-\nu_d^c(i, j))} \tag{2.20}$$

where

$$\mu_{d2}^c(i, j) = \frac{j \nu_d^c(i, j)}{1 - \Phi_2(-\nu_d^c(i, j))}. \tag{2.21}$$

Considering the two-tier overlaying structure in cellular/WLAN interworking, new and handoff data calls in the cellular-only area are prioritized by bandwidth reservation with the randomized guard channel policy. In this case, we use the following average departure rate to simplify analysis

$$\tilde{\mu}_d^c(i, j) = p_{d1}^c(j) \tilde{\mu}_{d1}^c(i, j) + p_{d2}^c(j) \tilde{\mu}_{d2}^c(i, j) \tag{2.22}$$

where $p_{d1}^c(\cdot)$ and $p_{d2}^c(\cdot)$ are respectively the fractions of traffic requesting admission to the cell from the cellular-only area and from the double-coverage area, given by

$$p_{d1}^c(j) = \lambda_{d1}^c(j)/\lambda_d^c(j), \quad p_{d2}^c(j) = \lambda_{d2}^c(j)/\lambda_d^c(j), \quad \lambda_d^c(j) = \lambda_{d1}^c(j) + \lambda_{d2}^c(j)$$

$$\lambda_{d1}^c(j) = \begin{cases} \lambda_{d1} + \lambda_{hd}^{cc} + \lambda_{hd}^{wc}, & j \leq \lfloor N_{d1}^c \rfloor, \quad N_{d1}^c = N_d^c - G_{d1}^c \\ \lambda_{d1}(N_{d1}^c - \lfloor N_{d1}^c \rfloor) + \lambda_{hd}^{cc} + \lambda_{hd}^{wc}, & j = \lfloor N_{d1}^c \rfloor + 1 \\ \lambda_{hd}^{cc} + \lambda_{hd}^{wc}, & \lfloor N_{d1}^c \rfloor + 2 \leq j \leq N_d^c \end{cases}$$

$$\lambda_{d2}^c(j) = \begin{cases} \lambda_{nd2}^c, & j \leq \lfloor N_{d2}^c \rfloor, \quad N_{d2}^c = N_d^c - G_{d2}^c \\ \lambda_{nd2}^c(N_{d2}^c - \lfloor N_{d2}^c \rfloor), & j = \lfloor N_{d2}^c \rfloor + 1 \\ 0, & \lfloor N_{d2}^c \rfloor + 2 \leq j \leq N_d^c. \end{cases}$$

Under the assumption of nearly complete decomposition of data traffic from voice, when there are i voice calls carried by the cell, the cell operates like a symmetric queue [20] for data with

$$\phi(j) = \tilde{\mu}_d^c(i, j), \ \gamma(l, j) = \delta(l, j) = \frac{1}{j}, \quad l = 1, 2, \ldots, j, \ j = 1, 2, \ldots, N_d^c$$

$$(2.23)$$

where $\phi(j)$ ($\phi(j) > 0$ if $j > 0$) is the total service rate when there are j customers (data calls) in the queue in positions $1, 2, \ldots, j$; $\gamma(l, j)$ is the fraction of the service rate directed to the customer in position l ($\sum_{l=1}^{j} \gamma(l, j) = 1$); $\delta(l, j+1) = \gamma(l, j+1)$ (symmetric condition) is the probability that an arriving customer moves into position l. A data call carried by the cell may depart due to a handoff to another cell or WLAN. This departure is independent of the queuing position of the data call and behaves like a multi-server loss system without waiting room. In addition, a data call may also depart from the cell due to call completion. Since all the bandwidth unused by current voice calls is shared equally by existing data calls in a PS manner, a fair share of the total service rate is dedicated to each data call irrelevant to its queueing position. Thus, a data call completion or arrival affects the amount of resources allocated to each data call, but each data call still keeps a fair share. Therefore, $\delta(l, j)$ and $\gamma(l, j)$ are independent of the queueing positions (i.e., l) of data calls and satisfy the symmetric condition. As a result, for data service, the cell can be modeled by a symmetric queue, which operates in a manner given by (2.23) and has a service capacity $(C^c - i R_v^c)$ with a probability $\pi_v^c(i)$, $i = 0, 1, \ldots, N_v^c$.

For symmetric queues such as processor-sharing queues and multi-server queues without waiting room (i.e., loss systems), a product-form stationary queue occupancy distribution exists and is applicable to arbitrarily distributed service requirements [20]. Then, the steady-state probability of j ($j = 1, 2, \ldots, N_d^c$) data calls in the cell is approximately given by

$$\pi_d^c(j) = \sum_{i=0}^{N_v^c} \left[\pi_v^c(i) \pi_d^c(0) \prod_{l=1}^{j} \frac{\lambda_d^c(l)}{\phi(l)} \right] = \sum_{i=0}^{N_v^c} \left[\pi_v^c(i) \pi_d^c(0) \prod_{l=1}^{j} \frac{\lambda_d^c(l)}{\tilde{\mu}_d^c(i, l)} \right] \quad (2.24)$$

where $\pi_d^c(0) = 1/C_3$. Then, the data call blocking and dropping probabilities in the cell can be obtained by replacing N_v^c, G_{v1}^c, G_{v2}^c, N_{v1}^c, N_{v2}^c and π_v^c in (2.11)–(2.13) with N_d^c, G_{d1}^c, G_{d2}^c, N_{d1}^c, N_{d2}^c and π_d^c, respectively.

On the other hand, the departure rate of data calls with i voice calls and j data calls in the WLAN is $j[\chi_d^w(i, j) + \eta^w]$. Similar to the derivation of $\pi_d^c(j)$, the steady-state probability of j data calls carried by the WLAN is given by

$$\pi_d^w(j) = \sum_{i=0}^{N_v^w} \left[\pi_v^w(i) \pi_d^w(0) \prod_{l=1}^{j} \frac{\lambda_d^w}{l[\chi_d^w(i, l) + \eta^w]} \right],$$

$$\pi_d^w(0) = 1/C_4, \ j = 1, 2, \ldots, N_d^w. \quad (2.25)$$

The data call blocking probability of the WLAN is then obtained as $B_d^w = \pi_d^w(N_d^w)$.

Table 2.1 System parameters

Parameter	Value	Parameter	Value
C^w	11 Mbit/s	C^c	2 Mbit/s
λ_{v1}	0.12 calls/s	λ_{v2}	0.18 calls/s
$(\mu_v)^{-1}$	140 s	R_v^c	12.2 kbit/s
$\tilde{B}_v(\tilde{B}_d)$	0.01	$\tilde{D}_v(\tilde{D}_d)$	0.001
f_d	64 KB	$\tilde{T}_d$	4 s
Δ	0.1	V_{lh}	0.6
$(\eta^c)^{-1}$	10 min	$(\eta^w)^{-1}$	14 min
p^{cc}	0.76	p^{cw}	0.24

As seen from (2.14), (2.16), (2.24), and (2.25), the mean data transfer time depends on the mean arrival rates and blocking and dropping probabilities of data calls, which are inter-dependent and need to be evaluated recursively.

2.4 Numerical Results and Discussion

Given in Table 2.1 are the system parameters, which are selected based on popularly deployed cellular networks (e.g., cdma2000) and WLAN standards (e.g., IEEE 802.11b). Applying the QoS evaluation approach in a search algorithm given in Table 2.2, we can obtain the voice and data allocation parameters. The best configuration should maximize the traffic load acceptable to a given cell/WLAN cluster. In the following, we discuss some important observations obtained from the numerical results of the searching process.

2.4.1 Accuracy Validation

We use a discrete event-driven simulator written in C/C++ language to verify the accuracy of our analysis. More than 10^7 voice and data call arrivals, departures and handoffs are generated in each simulation round to collect statistics on call blocking/dropping probabilities and data call transfer time. The results of multiple simulation rounds are averaged to remove randomness effect. The statistics are collected after the simulated system attains the equilibrium state.

Figures 2.3 and 2.4 illustrate the call-level QoS performance within the derived admission regions. As shown in Fig. 2.3, the simulation results of voice call blocking and dropping probabilities are very close to the analytical results and tightly bounded by the corresponding requirements. The performance fluctuation of handoff dropping probability is because the maximum numbers of calls allowed in the cell and WLAN are both integer variables. As we apply randomized guard channel policy to increase

Table 2.2 Search algorithm for allocation parameters

1: Derive cell capacity region of vectors (n_v^c, n_d^c) to satisfy $\frac{E_b}{N_0}$ requirements
2: Derive WLAN capacity region of vectors (n_v^w, n_d^w) to meet stability constraints
3: $N_{v,max}^w = \max(n_v^w)$: $(n_v^w, n_d^w) \in$ WLAN capacity region
4: $N_{v,max}^c = \max(n_v^c)$: $(n_v^c, n_d^c) \in$ cell capacity region
5: **for** $N_v^w = 0, ..., N_{v,max}^w$ **do**   // Evaluation for voice traffic
6:    By bisection search, determine $(N_v^c, G_{v1}^c, G_{v2}^c)$ s.t.
7:       $B_v^w B_{v2}^c \leq \tilde{B}_v$, $B_{v1}^c \leq \tilde{B}_v$, and $D_v^c \leq \tilde{D}_v$
8:    $N_{d,max}^w = \max(n_d^w)$ with $n_v^w = N_v^w$
9:    **for** $N_d^w = 0, ..., N_{d,max}^w$ **do**   // Evaluation for data traffic
10:       Initialize $\lambda_{d,min}$ and $\lambda_{d,max}$
11:       $\lambda_d \leftarrow (\lambda_{d,min} + \lambda_{d,max})/2$ // Update mean arrival rate of data calls λ_d
12:       By bisection search, determine $(N_d^c, G_{d1}^c, G_{d2}^c)$ and acceptable λ_d s.t.
13:          $B_d^w B_{d2}^c \leq \tilde{B}_d$, $B_{d1}^c \tilde{B}_d$, $D_d^c \leq \tilde{D}_d$, $\mathrm{E}[T_d^c] \leq \tilde{T}_d$, and $\mathrm{E}[T_d^w] \leq \tilde{T}_d$
14:       **if** solutions for $(N_d^c, G_{d1}^c, G_{d2}^c)$ exist **then**
15:          $\lambda_{d,min} \leftarrow \lambda_d$; $\lambda_d \leftarrow (\lambda_{d,min} + \lambda_{d,max})/2$
16:       **else**
17:          $\lambda_{d,max} \leftarrow \lambda_d$; $\lambda_d \leftarrow (\lambda_{d,min} + \lambda_{d,max})/2$
18:       **end if**
19:       **if** acceptable λ_d converges **then**
20:          break
21:       **end if**
22:       Record maximum acceptable λ_d
23:    **end for**
24: **end for**
25: Output (N_v^w, N_d^w), $(N_v^c, G_{v1}^c, G_{v2}^c)$, and $(N_d^c, G_{d1}^c, G_{d2}^c)$ that maximize acceptable λ_d

the granularity of bandwidth reservation, the fluctuation is actually smaller than that of traditional guard channel policy. On the other hand, as shown in Fig. 2.4, the mean data transfer time ($\mathrm{E}[T_d]$) is also well bounded and agrees well with the analytical results. To verify whether the user QoS is tightly bounded, we increase the maximum data call arrival rate λ_d obtained analytically by 1, 2, and 3 %, respectively. It is found that this increase results in QoS violation to the mean data transfer time. This indicates that the relative analytical error of the mean data transfer time is restricted within 1–3 %, and the upper bound of $\mathrm{E}[T_d]$ for the derivation of admission regions is tight in the case of integrated voice/data services.

2.4.2 Variation with WLAN Data Traffic

Figure 2.5 illustrates how the acceptable data traffic load (mean data call arrival rate λ_d) varies with the maximum number of data calls allowed in the WLAN (N_d^w) when the maximum number of voice calls allowed in the WLAN (N_v^w) is fixed to different values. As observed in Fig. 2.5a, the data traffic load increases with N_d^w when N_d^w is relatively small. This can be explained as follows. Due to the coupling between the

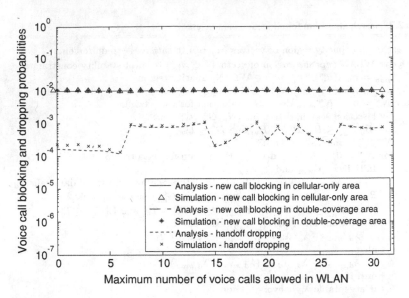

Fig. 2.3 Analytical and simulation results of voice call blocking/dropping probabilities

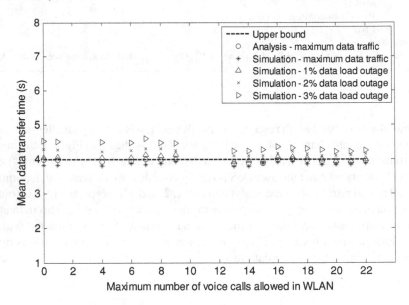

Fig. 2.4 Analytical and simulation results of mean data transfer time

cell and its overlaying WLAN, the less data calls allowed in the WLAN, the more data calls that need to be accommodated by the cell. With the PS sharing for data traffic, the less data calls admitted, the faster they will leave the system as a larger bandwidth is available for each data call. Due to user mobility, both the times that

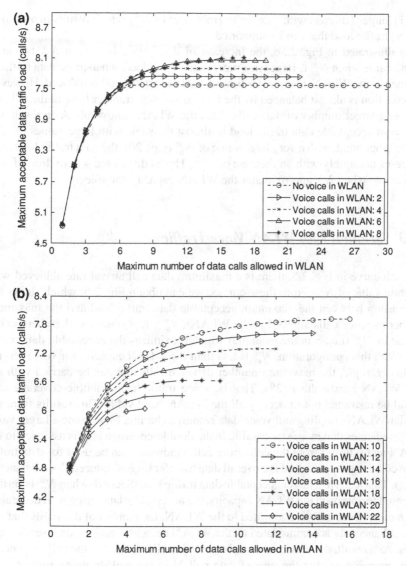

Fig. 2.5 Maximum acceptable data traffic load (mean data call arrival rate λ_d) versus maximum number of data calls allowed in the WLAN (N_d^w) under QoS constraints (blocking probabilities ≤ 0.01, dropping probabilities ≤ 0.001, and data transfer time ≤ 4 s)

a data call is carried by WLANs and by cells contribute to the total transfer time of a data call. Since the cell bandwidth is much lower than the WLAN bandwidth, the increase of data transfer time in cells cannot be compensated by the reduction of data transfer time when a data call is carried by WLANs. Hence, the mean data transfer time is longer (shorter) with a decrease (increase) of N_d^w, which results in a smaller (larger) traffic load that can be supported.

As illustrated in Fig. 2.5b, the increase of data traffic load with N_d^w becomes unnoticeable when N_d^w is large (say, more than 10). Indeed, when more data traffic is assigned to the WLAN, the transfer time of data calls in the cell is reduced. However, the reduction is almost balanced by the increase of data transfer time in the WLAN because a larger number of data calls share the WLAN bandwidth. As a result, the maximum acceptable data traffic load is almost the same with large values of N_d^w. On the other hand, with a very large value of N_v^w (e.g., 20), the data traffic load even decreases negligibly with an increase of N_d^w. This is due to the severe drop of data service rate when N_v^w approximates the WLAN capacity for voice.

2.4.3 Variation with WLAN Voice Traffic

For each curve in Fig. 2.5, there is a maximum data call arrival rate achieved with a certain value of N_d^w. From these curves, we can obtain Fig. 2.6, which shows the relationship between the maximum acceptable data traffic load and the maximum number of voice calls allowed in the WLAN (N_v^w). It is observed that there exists a value of N_v^w (i.e., 8 in the example) which maximizes the acceptable data traffic load. With this configuration, N_v^w is less than the WLAN capacity for voice service (in this example, the maximum number of voice calls that can be carried with the total WLAN bandwidth is 28). That is, voice traffic in the double-coverage area should be restricted not to occupy all the WLAN bandwidth. This results from the cellular/WLAN coupling and voice/data resource sharing. First, since a larger value of N_v^w indicates that more voice traffic in the double-coverage area is assigned to the WLAN and relieved from the cell, more cell bandwidth can be used for data traffic in the cellular-only area, and the overall data transfer time is reduced (load balancing effect). This leads to a larger acceptable data traffic load. Second, when N_v^w is further increased to approach the WLAN capacity, the acceptable data traffic load decreases. When more voice calls are admitted to the WLAN, the number of data calls that can be simultaneously accommodated by the WLAN decreases and the data service rate drops. As a result, the maximum number of data calls allowed in the cell (N_d^c) needs to be increased so that the overall data call blocking and dropping probabilities meet the corresponding requirements. Due to the much smaller cell bandwidth, an increased traffic load assigned to the cell results in a longer data transfer time. When the penalty incurred by voice support in WLANs overwhelms the advantage of the load balancing effect, the maximum acceptable data traffic load begins to decrease.

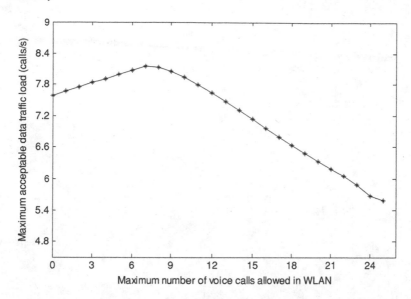

Fig. 2.6 Maximum acceptable data traffic load (mean data call arrival rate λ_d) versus maximum number of voice calls allowed in the WLAN (N_v^w) under QoS constraints (blocking probabilities ≤ 0.01, dropping probabilities ≤ 0.001, and data transfer time ≤ 4 s)

2.5 Summary

In this chapter, we introduce the *WLAN-first* resource allocation scheme for the cellular/WLAN intergraded network. In this simple and easy-to-implement scheme, WLANs are always preferred by all services whenever the WLAN access is available, so as to take advantage of the low cost and large bandwidth of WLANs. The main features and observations of this work are as follows:

- In this resource allocation scheme, new and handoff calls in different areas are prioritized with limited fractional guard channel policies. To compensate for the limited QoS differentiation capability of WLANs, restricted access mechanism is applied in the cell for resource sharing between voice and data services.
- An analytical model based on two-dimensional Markov processes is developed to evaluate QoS metrics in terms of call blocking/dropping probabilities and mean data transfer time. The model properly captures the location-dependent user mobility within the cell/WLAN cluster.
- Numerical results demonstrate that the QoS performance is closely related to the admission regions for voice and data services in the cell and WLAN. The best admission regions can be determined by applying the QoS evaluation in a search algorithm. Because data traffic accepts elastic bandwidth and better exploits the low mobility and large bandwidth in the double-coverage area, the maximum number of voice calls allowed in a WLAN should be restricted not to occupy all WLAN capacity.

Chapter 3
Call Admission Control with Randomized Access Selection

In Chap. 2, we present the WLAN-first allocation scheme for integrated voice and data services. The admission parameters are determined in such a way to maximize the overall resource utilization. Actually, the rationale behind is to properly distribute the multi-service traffic load to the integrated cell and WLAN so as to effectively exploit their complementary strength. In this chapter, we introduce a general admission control scheme with randomized access selection to enable distributed implementation. A more effective analytical approach is developed for QoS evaluation by means of moment generating functions (MGFs). Based on the analytical model, we further demonstrate the impact of mobility and traffic variability on the determination of admission parameters.

3.1 Distributed Admission Control

With the heterogeneous QoS support of the underlying integrated network, voice and data traffic in the double-coverage area should be properly directed to the cell and the WLAN. Due to the heterogeneity, especially when the two systems are loosely coupled, it is challenging for a central controller to timely obtain updated information of both systems (e.g., the numbers of ongoing calls in the cell and overlay WLANs) to make an optimal decision for each admission request. Also, there is a high control overhead since signaling messages have to traverse a long path involving many network elements. To reduce signaling overhead, frequent information exchanges may not be affordable; but outdated network information is adverse to decision accuracy. Consequently, in a loosely coupled cellular/WLAN network, distributed access selection and call admission control is more practical, although the decision may not be optimal in terms of maximizing resource utilization with QoS guarantee.

Instead of applying complex criteria for access selection, we consider a simple admission scheme to investigate the dependence of resource utilization on admission control and the impact of mobility and traffic characteristics. An incoming

W. Song and W. Zhuang, *Interworking of Wireless LANs and Cellular Networks*,
SpringerBriefs in Computer Science, DOI: 10.1007/978-1-4614-4379-7_3,

voice (data) call in the double-coverage area requests admission to the cell with a probability θ_v^c (θ_d^c), while it requests admission to the WLAN with a probability $\theta_v^w = 1 - \theta_v^c$ ($\theta_d^w = 1 - \theta_d^c$). The admission parameters θ_v^c and θ_d^c (or θ_v^w and θ_d^w) are determined for a given traffic load and broadcast to the associated mobiles. Then, a mobile can make a decision on its own according to these parameters and send the admission request to the corresponding target network. With the simplicity, this admission scheme can be implemented in a distributed manner. Also, as the network elements involved in the admission decision are limited to be as few as possible, the signaling overhead is reduced. The cellular network and the integrated WLANs only need to exchange information and update the above admission parameters with traffic variation. It is especially suited for the cases that it is not affordable to base each admission decision on the system states of both networks. By controlling the admission probabilities, the incoming traffic load is properly shared by the integrated cell and WLAN. This scheme enables simple implementation, although it may not fully exploit the performance gain achievable from the interworking.

This access selection strategy can be extended and applied in combination with other call admission criteria. For example, the cellular network also differs from WLANs in pricing rates. The service cost in the WLAN is generally much lower than that in the cellular network. A mobile user may prefer to get admitted to the WLAN for the low cost or to the cellular network if the service quality is more important. Suppose that an incoming call requests admission to the WLAN with a probability γ^w, while it requests admission to the cell with a probability $\gamma^c = 1 - \gamma^w$. Let ω_i, $i = 1, 2, ..., r$, denote the relative weights of r different criteria and θ_i^w (θ_i^c) the probability of selecting the WLAN (cell) as the target when the ith criterion is considered. Then, we have

$$\gamma^w = \sum_{i=1}^r \omega_i \theta_i^w, \quad \gamma^c = \sum_{i=1}^r \omega_i \theta_i^c \tag{3.1}$$

where

$$\sum_{i=1}^r \omega_i = 1, \quad \theta_i^w = 1 - \theta_i^c, \quad i = 1, 2, ..., r.$$

3.2 Determination of Admission Parameters

Similar to the WLAN-first scheme discussed in Chap. 2, the admission parameters θ_v^w and θ_d^w (or corresponding θ_v^c and θ_d^c) are determined to properly distribute the voice and data traffic load to the overlay cell and WLAN. First, the voice traffic load should be measured and estimated, as voice calls fluctuate in a larger time scale. Given the voice traffic load, the admission parameters can then be determined to maximize the acceptable data traffic load (λ_d) with QoS satisfaction. That is,

$$\max_{(\theta_v^w, \theta_d^w)} \lambda_d, \quad s.t.$$

$$\theta_v^w B_v^w + \theta_v^c B_{v2}^c \leq \tilde{B}_v, \quad B_{v1}^c \leq \tilde{B}_v, \quad D_v^c \leq \tilde{D}_v$$

$$\theta_d^w B_d^w + \theta_d^c B_{d2}^c \leq \tilde{B}_d, \quad B_{d1}^c \leq \tilde{B}_d, \quad D_d^c \leq \tilde{D}_d, \quad \mathrm{E}[T_d^c] \leq \tilde{T}_d, \quad \mathrm{E}[T_d^w] \leq \tilde{T}_d$$

$$(3.2)$$

where $\theta_v^w B_v^w + \theta_v^c B_{v2}^c$ is the blocking probability in the double-coverage area for new voice calls, and $\theta_d^w B_d^w + \theta_d^c B_{d2}^c$ is that for new data calls. The analytical model proposed in Chap. 2 is also applicable to the QoS evaluation for this case.

The mean arrival rates of new voice and data calls to the cell from the double-coverage area are respectively

$$\lambda_{nv2}^c = \theta_v^c \lambda_{v2}, \quad \lambda_{nd2}^c = \theta_d^c \lambda_{d2}. \tag{3.3}$$

Likewise, the mean arrival rates of new voice and data calls to the WLAN from the double-coverage area are respectively

$$\lambda_{nv}^w = \theta_v^w \lambda_{v2}, \quad \lambda_{nd}^w = \theta_d^w \lambda_{d2}. \tag{3.4}$$

Nonetheless, as the QoS evaluation for the cell is based on two-dimensional Markov processes, the computation complexity increases with the size of state space. In this work, we circumvent the computation complexity of solving large-scale balance equations by means of moment generating functions (MGFs).

3.2.1 Performance Analysis for the WLAN

The voice traffic load offered to the WLAN includes new calls from the double-coverage area with a mean rate λ_{nv}^w and handoff calls from the overlaying cell with a mean rate λ_{hv}^{cw}. Assume that the user residence time within the double-coverage area (T_r^w) is modelled by a hyper-exponential distribution in (1.3). With Poisson arrivals and exponential voice call duration (T_v), the channel holding time $\min(T_v, T_r^w)$ is not exponential and the voice calls in the WLAN can be modeled by an $M/G/K/K$ queue. With the insensitive property of $M/G/K/K$ queues [20], the steady-state probability of voice calls in the WLAN is given by

$$\pi_v^w(k) = \pi_v^w(0) \prod_{i=1}^{k} \frac{\lambda_v^w(i)}{i \mu_v^w}, \quad k = 1, \ldots, N_v^w \tag{3.5}$$

where

$$E[\min(T_v, T_r^w)] = \frac{a}{a+1} \cdot \frac{1}{a\eta^w + \mu_v} + \frac{1}{a+1} \cdot \frac{1}{\frac{1}{a}\eta^w + \mu_v} \triangleq \frac{1}{\mu_v^w}$$

$$\lambda_v^w(i) = \begin{cases} \lambda_{nv}^w + \lambda_{hv}^{cw}, & i \le M_v^w \\ \lambda_{nv}^w, & M_v^w + 1 \le i \le N_v^w. \end{cases}$$

Within the two-tier overlaying structure, the vertical handoff from the cell to the overlaying WLAN is not necessary but optional to maintain an ongoing call. Hence, the handoff traffic load to the WLAN can be controlled by properly adjusting the admission parameters of the WLAN, e.g., by using a simple guard channel policy. Let G_v^w (G_d^w) denote the number of guard channels reserved in the WLAN for new voice (data) calls from the double-coverage area. Then, a handoff voice (data) call from the cell is admitted to the WLAN if the number of voice (data) calls in the WLAN is less than $M_v^w = N_v^w - G_v^w$ ($M_d^w = N_d^w - G_d^w$) and rejected otherwise. Hence, the voice call blocking probability of the WLAN is $B_v^w = \pi_v^w(N_v^w)$, and the rejection probability for handoff voice calls from the cell is $D_v^w = \sum_{i=M_v^w}^{N_v^w} \pi_v^w(i)$.

The data call blocking probability and rejection probability of the WLAN can be evaluated similarly except for a state-dependent service rate. With i voice calls and j data calls admitted in the WLAN, the average channel holding time of data calls of the two virtual classes is respectively

$$\frac{1}{\mu_{d1}^w(i,j)} = \frac{a}{a+1} \cdot \frac{1}{a\eta^w + \frac{R_d^w(i,j)}{f_d/b}} + \frac{1}{a+1} \cdot \frac{1}{\frac{1}{a}\eta^w + \frac{R_d^w(i,j)}{f_d/b}} \tag{3.6}$$

$$\frac{1}{\mu_{d2}^w(i,j)} = \frac{a}{a+1} \cdot \frac{1}{a\eta^w + \frac{R_d^w(i,j)}{bf_d}} + \frac{1}{a+1} \cdot \frac{1}{\frac{1}{a}\eta^w + \frac{R_d^w(i,j)}{bf_d}} \tag{3.7}$$

where $R_d^w(\cdot)$ is the average transmission rate (in bps) for packets from a data call. Given that the data call size follows the hyper-exponential distribution in (1.2), the data traffic to the WLAN can be viewed as two virtual classes [27] with Poisson arrival rates $\frac{b}{b+1}\lambda_d^w(j)$ and $\frac{1}{b+1}\lambda_d^w(j)$, respectively, and exponentially distributed service requirements with mean f_d/b and bf_d, respectively, where $\lambda_d^w(j) = \lambda_{nd}^w + \lambda_{hd}^{cw}$ when $j \le M_d^w$, and $\lambda_d^w(j) = \lambda_{nd}^w$ when $M_d^w + 1 \le j \le N_d^w$. Then, the steady-state probability of data calls in the WLAN is approximated by

$$\pi_d^w(j) = \sum_{i=0}^{N_v^w} \pi_v^w(i)\tilde{\pi}_d^w(j|i) \tag{3.8}$$

where

$$\tilde{\pi}_d^w(j|i) = \tilde{\pi}_d^w(0|i) \prod_{l=1}^{j} \left[\frac{\frac{b}{b+1}\lambda_d^w(l)}{l\mu_{d1}^w(i,l)} + \frac{\frac{1}{b+1}\lambda_d^w(l)}{l\mu_{d2}^w(i,l)} \right], \qquad j = 1, \ldots, N_d^w. \tag{3.9}$$

Similarly, the data call blocking probability of the WLAN is obtained as $B_d^w = \pi_d^w$ (N_d^w), and the rejection probability for handoff data calls from the cell is $D_d^w = \sum_{i=M_d^w}^{N_d^w} \pi_d^w(i)$. From the Little's law, the mean data transfer time in the WLAN is given by

$$\mathrm{E}[T_d^w] = \sum_{i=0}^{N_v^w} \pi_v^w(i) \frac{\sum_{l=1}^{N_d^w} l \tilde{\pi}_d^w(l|i)}{(1 - B_d^w)\lambda_{nd}^w + (1 - D_d^w)\lambda_{hd}^{cw}}. \qquad (3.10)$$

3.2.2 Performance Analysis for the Cell

Due to the location-dependent mobility within a cell, calls in the cellular-only area and the double-coverage area differ in channel holding time. Depending on the WLAN state, the average channel holding time of voice calls ($1/\mu_{v1}^c$ and $1/\mu_{v2}^c$) can be derived from (2.6) and (2.7) for the cellular-only area and double-coverage area, respectively. To simplify analysis, we take an average for the mean service rates of voice calls in the cellular-only area and the double-coverage area, which are respectively given by

$$\tilde{\mu}_{v1}^c = D_v^w \mu_{v1}^c + (1 - D_v^w)(\mu_v + \eta^c) \qquad (3.11)$$

$$\tilde{\mu}_{v2}^c = D_v^w \mu_{v2}^c + (1 - D_v^w)\frac{\mu_v}{1 - \psi(-\mu_v)}. \qquad (3.12)$$

That is, voice calls admitted to the cell from the cellular-only area and the double-coverage area follow average channel holding time $(\tilde{\mu}_{v1}^c)^{-1}$ and $(\tilde{\mu}_{v2}^c)^{-1}$, respectively. As discussed in Sect. 2.2, we can use the recursive method in [55] to approximate the probability of having k voice calls in the cell as

$$\pi_v^c(k) = \pi_v^c(0) \prod_{i=1}^{k} \left[\frac{\lambda_{v1}^c(i)}{i\tilde{\mu}_{v1}^c} + \frac{\lambda_{v2}^c(i)}{i\tilde{\mu}_{v2}^c} \right], \quad k = 1, \ldots, N_v^c \qquad (3.13)$$

where

$$\lambda_{v1}^c(i) = \begin{cases} \lambda_{v1} + \lambda_{hv}^{cc} + \lambda_{hv}^{wc}, & i \leq \lfloor N_v^c - G_{v1}^c \rfloor \\ [1 - (G_{v1}^c - \lfloor G_{v1}^c \rfloor)]\lambda_{v1} + \lambda_{hv}^{cc} + \lambda_{hv}^{wc}, & i = \lceil N_v^c - G_{v1}^c \rceil \\ \lambda_{hv}^{cc} + \lambda_{hv}^{wc}, & \lceil N_v^c - G_{v1}^c \rceil + 1 \leq i \leq N_v^c \end{cases}$$

$$\lambda_{v2}^c(i) = \begin{cases} \lambda_{nv2}^c, & i \leq \lfloor N_v^c - G_{v2}^c \rfloor \\ [1 - (G_{v1}^c - \lfloor G_{v1}^c \rfloor)]\lambda_{nv2}^c, & i = \lceil N_v^c - G_{v2}^c \rceil. \end{cases}$$

The voice call blocking and dropping probabilities of the cell can then be obtained from $\pi_v^c(k)$, $k = 0, 1, \ldots, N_v^c$.

To evaluate the QoS metrics of data calls in the cell, two other important aspects need to be properly addressed. First, with the restricted access mechanism, the data

call service rates become dependent on both voice and data calls in the cell, as all
bandwidth unused by current voice traffic is shared equally by active data calls.
Second, the high variability of data call size should be properly addressed in the QoS
evaluation. Since the remaining bandwidth unused by current voice calls is shared
equally by existing data calls in a processor sharing (PS) manner, a fair share of the
total service rate is dedicated to each data call.

Data calls admitted in the cell are differentiated into two virtual classes with exponentially distributed service requirements with mean $\frac{1}{b}f_d$ and bf_d, respectively [27].
Then, data service in the cell is modeled by a symmetric queue serving multiple
classes. Given i voice calls and j data calls in the cell, as in (3.11) and (3.12), the
service rates of the two virtual classes of data calls in the cellular-only area can be
approximated by

$$\tilde{\mu}_{d1}^{c1}(i, j) = D_d^w \mu_{d1}^{c1}(i, j) + (1 - D_d^w)\left[b\nu_d^c(i, j) + \eta^c\right] \tag{3.14}$$

$$\tilde{\mu}_{d2}^{c1}(i, j) = D_d^w \mu_{d2}^{c1}(i, j) + (1 - D_d^w)\left[\nu_d^c(i, j)/b + \eta^c\right] \tag{3.15}$$

where $\nu_d^c(i, j) = \frac{C^c - iR_v^c}{jf_d}$ and

$$\mu_{d1}^{c1}(i, j) = \frac{b\nu_d^c(i, j)}{1 - \Phi_1(-b\nu_d^c(i, j))}, \qquad \mu_{d2}^{c1}(i, j) = \frac{\nu_d^c(i, j)/b}{1 - \Phi_1(-\nu_d^c(i, j)/b)}. \tag{3.16}$$

Similarly, the service rates of the two virtual classes of data calls admitted to the cell
from the double-coverage area can be obtained as

$$\tilde{\mu}_{d1}^{c2}(i, j) = D_d^w \mu_{d1}^{c2}(i, j) + (1 - D_d^w)\frac{b\nu_d^c(i, j)}{1 - \psi(-b\nu_d^c(i, j))} \tag{3.17}$$

$$\tilde{\mu}_{d2}^{c2}(i, j) = D_d^w \mu_{d2}^{c2}(i, j) + (1 - D_d^w)\frac{\nu_d^c(i, j)/b}{1 - \psi(-\nu_d^c(i, j)/b)} \tag{3.18}$$

where

$$\mu_{d1}^{c2}(i, j) = \frac{b\nu_d^c(i, j)}{1 - \Phi_2(-b\nu_d^c(i, j))}, \qquad \mu_{d2}^{c2}(i, j) = \frac{\nu_d^c(i, j)/b}{1 - \Phi_2(\nu_d^c(i, j)/b)}. \tag{3.19}$$

For symmetric queues such as processor-sharing queues and multi-server queues
without waiting room (i.e., loss systems), a product-form stationary queue occupancy distribution exists and is applicable to arbitrarily distributed service requirements [20]. Hence, given i voice calls in the cell, the equilibrium distribution of the
symmetric queue for data traffic in the cell is given by

$$\pi_d^c(j|i) = \tilde{\pi}_d^c(0|i)\prod_{l=1}^{j}\left[\frac{\frac{b}{b+1}\lambda_{d1}^c(l)}{l\tilde{\mu}_{d1}^{c1}(i, l)} + \frac{\frac{1}{b+1}\lambda_{d1}^c(l)}{l\tilde{\mu}_{d2}^{c1}(i, l)} + \frac{\frac{b}{b+1}\lambda_{d2}^c(l)}{l\tilde{\mu}_{d1}^{c2}(i, l)} + \frac{\frac{1}{b+1}\lambda_{d2}^c(l)}{l\tilde{\mu}_{d2}^{c2}(i, l)}\right]$$
$$\tag{3.20}$$

where $\lambda_{d1}^c(\cdot)$ and $\lambda_{d2}^c(\cdot)$ are the mean arrival rates of data calls from the cellular-only area and the double-coverage area, respectively, given by

$$\lambda_{d1}^c(l) = \begin{cases} \lambda_{d1} + \lambda_{hd}^{cc} + \lambda_{hd}^{wc}, & l \le \lfloor N_d^c - G_{d1}^c \rfloor \\ [1 - (G_{d1}^c - \lfloor G_{d1}^c \rfloor)]\lambda_{d1} + \lambda_{hd}^{cc} + \lambda_{hd}^{wc}, & l = \lceil N_d^c - G_{d1}^c \rceil \\ \lambda_{hd}^{cc} + \lambda_{hd}^{wc}, & \lceil N_d^c - G_{d1}^c \rceil + 1 \le l \le N_d^c \end{cases}$$

$$\lambda_{d2}^c(l) = \begin{cases} \lambda_{nd2}^c, & l \le \lfloor N_d^c - G_{d2}^c \rfloor \\ [1 - (G_{d1}^c - \lfloor G_{d1}^c \rfloor)]\lambda_{nd2}^c, & l = \lceil N_d^c - G_{d2}^c \rceil. \end{cases}$$

Let $\pi_d^c(\cdot)$ denote the steady-state probability of data calls in the cell. Then,

$$\pi_d^c(j) = \sum_{i=0}^{N_v^c} \pi_v^c(i)\tilde{\pi}_d^c(j|i), \qquad j = 0, 1, \ldots, N_d^c. \tag{3.21}$$

The data call blocking and dropping probabilities and mean data transfer time can be obtained from π_d^c as in Chap. 2. From the Little's law, the mean data transfer time in the cell can be obtained as

$$E[T_d^c] = \sum_{i=0}^{N_v^c} \pi_v^c(i) \frac{\sum_{l=1}^{N_d^c} l\tilde{\pi}_d^c(l|i)}{\lambda_{d1}^c + (1 - B_d^c)\lambda_{nd2}^c}. \tag{3.22}$$

3.3 Numerical Results and Discussion

In this section, we first validate the accuracy of the QoS evaluation approaches based on Markov processes and MGFs. Further, we investigate the impact of traffic and mobility variability on the determination of admission parameters and resulting resource utilization. The system parameters are the same as those given in Table 2.1.

3.3.1 Accuracy Validation

In Chap. 2, we discuss a QoS evaluation approach based on two-dimensional Markov processes. The MGF approach introduced in this chapter evaluates the QoS metrics with closed-form approximation with a much lower complexity, which makes it possible to have the admission parameters adaptive to time-varying traffic load. Possible approximation errors may be introduced due to the location-dependent user mobility model within the cell, traffic prioritization by the limited fractional guard channel policies, and correlation between voice and data traffic. Here, we conduct computer simulations to verify the analytical results under different user mobility and traffic conditions.

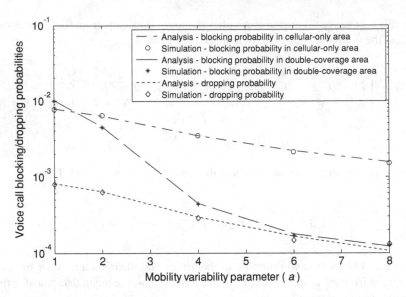

Fig. 3.1 Analytical and simulation results of voice call blocking/dropping probabilities

Figure 3.1 illustrates the results for voice call blocking and dropping probabilities. With a larger mobility variability parameter a, the user residence time in the WLAN deviates more from the exponential distribution and has a higher variability. As seen, the analytical results agree well with the simulation results for different values of a.

Figures 3.2 and 3.3 show the analytical and simulation results for data call QoS metrics such as data call blocking/dropping probabilities and mean data transfer time. It is observed that the analytical results are very close to the simulation results. The gap is much less than 10%, although it increases slightly with the data call variability parameter b for mean data transfer time in the cell ($E[T_d^c]$). The error is induced by the assumption of nearly complete decomposability to decouple the analysis for data calls from voice. To take advantage of the data traffic elasticity, data calls are served under the PS scheduling discipline. The mean data transfer time is insensitive to data call size distribution if the total service capacity is fixed. However, the insensitivity is generally lost with a varying capacity, and the call-level QoS improves with a higher variability for data call size [23]. With this admission scheme, the bandwidth available to data traffic actually fluctuates with voice call arrivals and departures. As a result, the mean data transfer time $E[T_d^c]$ is overestimated with a larger value of b. Nonetheless, the analysis is still quite accurate especially when data calls arrive and depart in a much smaller time scale than voice calls.

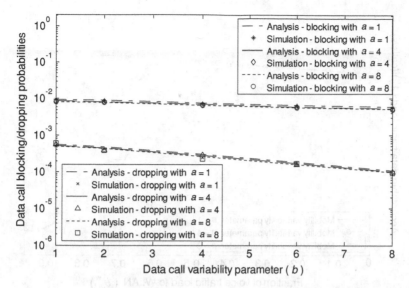

Fig. 3.2 Analytical and simulation results of data call blocking/dropping probabilities

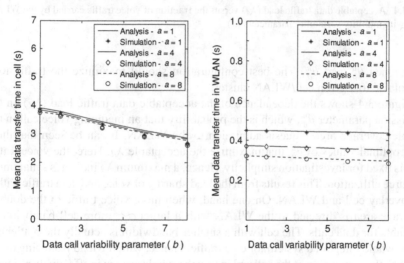

Fig. 3.3 Analytical and simulation results of mean data transfer time. **a** Mean transfer time of data calls in the cell ($E[T_d^c]$). **b** Mean transfer time of data calls in the WLAN ($E[T_d^w]$)

3.3.2 Dependence of Utilization on Admission Parameters

The MGF-based approach given in Sect. 3.2 evaluates QoS metrics accurately and effectively. The admission parameters θ_v^w and θ_d^w can be determined by applying the MGF-based QoS evaluation in a search algorithm similar to that given in Table 2.2. The corresponding voice and data admission regions can also be obtained for the

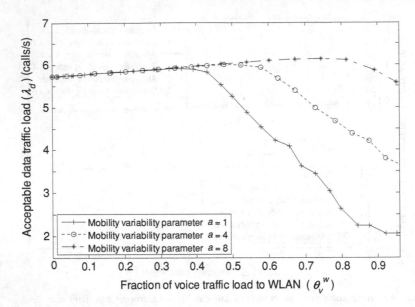

Fig. 3.4 Acceptable data traffic load (λ_d) versus the fraction of voice traffic carried by the WLAN (θ_v^w) with mobility variability parameters (a)

determined θ_v^w and θ_d^w. The best configuration should maximize the traffic load acceptable to a given cell/WLAN cluster.

Figure 3.4 shows the dependence of the acceptable data traffic load (λ_d) on the admission parameter θ_v^w, which is the probability that an incoming voice call in the double-coverage area requests admission to the WLAN. It can be seen that there exist optimal values of θ_v^w that maximize the acceptable λ_d. Here, the voice traffic load is fixed for investigation simplicity. Hence, a maximum λ_d indicates a maximum resource utilization. This results from the load sharing of voice and data traffic within the overlay cell and WLAN. On one hand, when more voice traffic in the double-coverage area is directed to the WLAN with a larger θ_v^w, more cell bandwidth is available for data calls. The cell with a smaller bandwidth is actually the bottleneck of the whole integrated system for data traffic. Hence, with the load balancing of the WLAN, the congestion of the cell and in turn the whole system is effectively relieved. On the other hand, with a larger θ_v^w, the maximum number of voice calls allowed in the WLAN (N_v^w) should also be larger to meet the voice call blocking/dropping probability requirements. However, the WLAN is very inefficient in supporting voice traffic. The small coverage of the WLAN also leads to frequent vertical handoffs between the cell and the WLAN, which may degrade voice quality and increase call dropping possibility. The break of a call into more service stages is also detrimental to multiplexing gain. Although the voice traffic load to the cell is reduced to an extent, the number of data calls that can be accommodated by the WLAN is also significantly reduced. As a result, the total acceptable traffic load starts to decrease when θ_v^w is larger than a threshold.

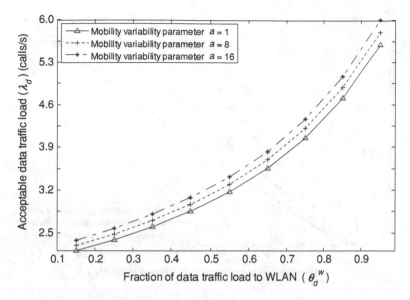

Fig. 3.5 Acceptable data traffic load (λ_d) versus the fraction of data traffic carried by the WLAN (θ_d^w) with mobility variability parameters (a)

Figure 3.5 shows the variation of the acceptable data traffic load (λ_d) with θ_d^w, i.e., the probability that a data call in the double-coverage area requests admission to the WLAN, which is correlated with θ_v^w. With a smaller θ_v^w to carry a less voice traffic load in the WLAN, θ_d^w can be larger to admit more data calls and provide enough bandwidth for each admitted call. In Fig. 3.4, before θ_v^w reaches the point for a maximum acceptable data traffic load, θ_v^w is less than 0.74, and N_v^w less than 16 is sufficient to meet the bounds for voice call blocking/dropping probabilities. For these cases, a high data call throughput is achievable since the number of voice calls allowed in the WLAN is quite restricted, and θ_d^w can be as large as 90 % to carry almost all the data traffic in the double-coverage area. Within this region, a larger θ_v^w alleviates more voice traffic from the cell but does not affect much the data service in the WLAN, which results in a larger acceptable λ_d. On the other hand, when θ_v^w and corresponding N_v^w are further increased, θ_d^w is even smaller and the WLAN cannot carry a large portion of the data traffic load in the double-coverage area. As a result, the bottleneck effect of the cell becomes evident. Thus, as shown in Fig. 3.5, the acceptable λ_d decreases with a smaller θ_d^w. In conclusion, the effectiveness of the WLAN as a complement to the cell may be greatly jeopardized with a small θ_d^w and a very large θ_v^w. To maximize the utilization, θ_v^w should be large enough to balance voice traffic load from the cell and also small enough to avoid an inefficient utilization of the WLAN for voice support.

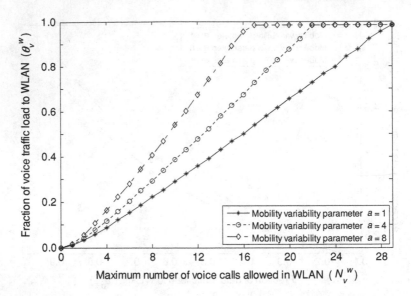

Fig. 3.6 Fraction of voice traffic carried by the WLAN (θ_v^w) versus the maximum number of voice calls allowed in the WLAN (N_v^w) with mobility variability parameters (a)

3.3.3 Impact of User Mobility and Traffic Variability

The curves in Figs. 3.4 and 3.5 are obtained with different mobility variability parameters (a). It is observed that the acceptable data traffic load (λ_d) is larger with a larger value of a. That is, a higher utilization is achievable when the variability of user mobility in the double-coverage area is higher. As illustrated in Fig. 3.4, when $a = 1$, 4, and 8, the highest utilization is achieved with $\theta_v^w = 0.36$, 0.48, and 0.74, respectively. A larger parameter a indicates that more users staying within the WLAN for a shorter time. Then, more voice calls may have a smaller channel holding time and occupy the WLAN bandwidth for a less time. As shown in Fig. 3.6, given a fixed N_v^w (i.e., the maximum number of voice calls allowed in the WLAN), when the parameter a is larger, a larger fraction of voice calls in the double-coverage area can be carried by the WLAN and relieved from the cell. Therefore, the data call throughput in the cell is higher and more traffic is acceptable with QoS satisfaction.

The data call variability also affects the admission parameters and resource utilization. As shown in Fig. 3.7, more data traffic is acceptable with a larger value of b, which indicates a higher variability of data call size. For a fixed mean of data call size, a larger value of b indicates that more data calls have a smaller size and less have an extremely larger size. Hence, more data calls have a shorter channel holding time and can be carried by the WLAN with a high throughput. Also, more data calls are likely to complete service within the WLAN and do not need to hand over to the cell when users move out of the WLAN coverage. Therefore, the WLAN bandwidth is more effectively utilized to reduce the traffic load to the cell of small

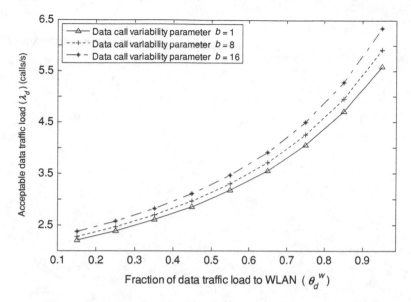

Fig. 3.7 Acceptable data traffic load (λ_d) versus the fraction of data traffic carried by the WLAN (θ_d^w) with data call variability parameters (b)

bandwidth. With the MGF-based approach, we can take into account the variability of user mobility and data traffic in determining the admission parameters.

3.4 Summary

In this chapter, we present an admission control scheme with randomized access selection for cellular/WLAN interworking. An incoming call in the double-coverage area is assigned to the overlay cell and WLAN according to properly defined admission probabilities. The main features and observations of this work are as follows:

- The randomized scheme enables distributed control to render feasible implementation in a loosely coupled network. The control overhead is reduced by avoiding frequent signaling exchanges to update states of both networks.
- An MGF-based analytical approach is developed to effectively and accurately evaluate the QoS metrics such as call blocking/dropping probabilities and mean data transfer time. The admission parameters are determined with the approach to achieve a high utilization.
- We demonstrate the impact of user mobility and data traffic variability on resource utilization. It is observed that the high data traffic variability can be exploited to improve the interworking effectiveness.

Chapter 4
Size-Based Load Sharing with SRPT Scheduling

From the preceding studies, we can see that traffic load sharing is essentially important to maximize the interworking effectiveness. As discussed in Chaps. 2 and 3, the
voice and data traffic load is distributed to the coupled systems via access selection
and call admission control, which properly differentiates upward/downward vertical handoff calls, horizontal handoff calls, new calls in the cellular-only area and
the double-coverage area. The impact of user mobility is also examined in Chap. 3.
Nonetheless, the research is focused on vertical handoff calls crossing WLAN bor-
ders as most previous works. Actually, access reselection can also be performed via
dynamic vertical handoff within the overlay area. In [22], there is some initial study
on dynamic session transfer in hierarchical integrated networks as an analogy to task
migration in distributed operating systems. However, little analytical work consid-
ers the dynamic vertical handoff within the overlay area triggered by network states
instead of user mobility. As the dynamics of both integrated systems are involved,
the load sharing problem becomes very complex, particularly for a multi-service
scenario. In this work, we present a size-based load sharing scheme for voice and
elastic data traffic in the cellular/WLAN integrated network.

4.1 Access Selection and Scheduling for Heavy-Tailed Data Calls

Due to contention-based access and excessive control overhead, it is very inefficient
to support real-time services in the WLAN. In contrast, the cellular network has
strength in real-time service provisioning. The large cell size and ubiquitous cellular
coverage can reduce handoff frequency and in turn the impact of handoff latency
on the delay-sensitive real-time traffic. Thus, an incoming voice call is preferably
distributed to the cell, and overflows to the WLAN only if there is not sufficient spare
capacity in the cell. The following solution further incorporates access reselection
for voice calls via dynamic vertical handoff from the WLAN to the cell, which
can be performed whenever the cell has spare capacity to accommodate more voice

W. Song and W. Zhuang, *Interworking of Wireless LANs and Cellular Networks*,   43
SpringerBriefs in Computer Science, DOI: 10.1007/978-1-4614-4379-7_4,
© The Author(s) 2012

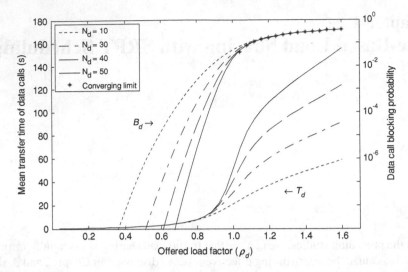

Fig. 4.1 Data call QoS in terms of mean transfer time (T_d) and data call blocking probability (B_d) under PS service discipline versus offered load factor (ρ_d) and number of admitted data calls (N_d)

calls. As such, voice calls are more concentrated in the cell and provisioned fine QoS guarantee. The bandwidth unused by voice traffic in the two systems can then be pooled to serve data calls. The rationale behind the idea can be understood by viewing the integrated cell and WLAN as two coupled queueing systems with service rates C_1 and C_2, respectively. By exploiting the cellular/WLAN interworking and vertical handoff, the performance of the two coupled systems within the overlay area can approach that of one queue with a larger service rate ($C_1 + C_2$), which maximizes the multiplexing gain [58].

On the other hand, admitted calls in the WLAN are served under the processor sharing (PS) scheduling with the contention-based access. The service queue with PS discipline exhibits unique characteristics, which should be considered in the resource management and may significantly affect the utilization. Based on queueing analysis for $M/G/1/K - PS$ queues, Fig. 4.1 is obtained to illustrate the dependence of QoS on the offered traffic load factor (ρ_d) and the number of admitted data calls (N_d). It can be seen that the mean transfer time T_d increases relatively slowly with ρ_d, when the system is underloaded with $\rho_d \ll 1$. For a moderately large value of N_d, the data call blocking probability B_d is very small and T_d is almost independent of N_d. However, when overload occurs with $\rho_d \geq 1$, T_d increases fast and almost linearly with ρ_d and N_d, while B_d converges fast to the limit $\frac{\rho_d-1}{\rho_d}$ with a moderately large value of N_d [14]. Hence, admitting more calls is not effective to reduce the blocking probability in overload but may significantly degrade the perceived performance. It is important to ensure that the system operates in a normal load condition, so that the blocking probability is bounded and a sufficiently high throughput is maintained for admitted calls [14].

Because voice calls are preferably distributed to the cell via initial call admission and access reselection via dynamic vertical handoff, the average cell bandwidth available to data traffic is relatively low when the voice traffic load is high. It is necessary and feasible to serve data calls in the cell with a more efficient service discipline. In this work, we consider the shortest remaining processing time (SRPT) scheduling discipline, which is optimal in terms of minimizing the mean transfer time. Under the SRPT, only one call with the least remaining data to transmit is scheduled first and receives service at an instant. Given an incoming data call with a size smaller than the remaining data size of the call in service, the ongoing call is preempted and waits in the queue, while the new call is served subsequently. In contrast, under the PS, each ongoing call shares an equal quantum of service. As such, smaller-size calls under the SRPT will not be stuck in the system for such a long duration as when the bandwidth is shared with data calls of a larger size.

It is known that the SRPT can significantly outperform the PS when the call size is heavy-tailed and the load is high. It may be suspected that the improvement of SRPT over PS comes at the expense of a longer transfer time for calls with a larger data size. Thus, the SRPT is often thought to be unfair as it favors short calls and penalizes long calls. An argument for this claim is the Kleinrock conservation law [21], which holds for service disciplines not making use of the size but is not necessarily true for size-based disciplines such as the SRPT. It is proved in [5] that, for any load condition and any continuous heavy-tailed size distribution with finite mean and variance, at least 99 % of the data calls have a smaller transfer time under the SRPT than under the PS. These 99 % of calls actually do significantly better, and the unfairness of SRPT diminishes with the heavy-tailed property. In addition, the control overhead of SRPT such as for preemption is also not higher than that of PS [5]. In practical systems, the PS may be implemented in a round-robin manner and each call is preempted after receiving one quantum of service. In contrast, the preemption of SRPT only occurs when a new call of a smaller size arrives, which involves less preemption overhead. As the SRPT is applied at the call level instead of the packet level, the implementation complexity and cost should be affordable.

Consider some specific elastic data applications such as Web browsing and file transfer. They usually preserve a request-response pattern and are primarily unidirectional from application servers to user terminals. The Web documents or data files are pre-stored in a Web server or file server. It is possible to know the data call size a priori from session signaling. For example, a session description protocol (SDP) offer/answer mechanism has been proposed as an Internet draft for file transfer [16]. By introducing a set of new SDP attributes, it is possible to deliver some meta information of the file (such as content type and size) before the actual transfer. On the other hand, cross-layer design has become very popular and essential in the wireless domain to address the unique challenges such as the scarce radio resources and highly error-prone transmission conditions. The information exchange across different protocol layers can further improve the system performance.

Hence, this load sharing scheme exploits the meta information of data calls that can be passed to the network layer. In particular, a data call is distributed to the cell if the call size is not greater than a threshold Ω_d and the cell bandwidth available to data traffic is at least R_d^c. Otherwise, that data call is assigned to the WLAN. The overall resource utilization can be improved without degrading the user QoS experience by properly selecting the size threshold (to be discussed in Sect. 4.2.2).

4.2 Performance Analysis

In this section, we present performance evaluation for voice/data call blocking probabilities and mean data transfer time, based on which the data call size threshold (Ω_d) can be determined.

4.2.1 Analytical Model for QoS Evaluation

As analyzed in [43], data calls in the WLAN share the available bandwidth in a PS manner. Under the PS, the mean transfer time is insensitive to the call size distribution if the overall service capacity is fixed. Nonetheless, due to the random access in the WLAN, the bandwidth available to data traffic actually fluctuates not only with voice call arrivals/departures but also with the contention status. The insensitivity of mean transfer time is generally lost in case of a varying capacity [23]. For data calls of a heavy-tailed size and high variability, the call-level performance even improves over the case with an exponentially distributed data call size. However, with admission control in place, the insensitivity can be retained for a high load condition, where proper resource allocation and load control are critical to prevent QoS violation. In a light load case, the call blocking probability is usually sufficiently low and all admitted calls are provided satisfactory QoS. Hence, we assume that the QoS of data traffic in the WLAN is insensitive to the heavy-tailed call size distribution. The insensitivity assumption is validated by the numerical results in Sect. 4.3, although conservative control is likely for a light load due to QoS underestimation.

Assume that voice and data call arrivals to the double-coverage area are independent Poisson processes with mean rates denoted by λ_v and λ_d, respectively. Since data calls are assigned to the integrated cell and WLAN based on the data call size and bandwidth occupancy, the data call arrivals to the cell and the WLAN are still Poisson processes with mean rates denoted by λ_d^c and λ_d^w, respectively. Given the insensitivity assumption for data service in the WLAN, we can model the integrated cell/WLAN cluster with a three-dimensional Markov process, in which the state (i, j, k) denotes the numbers of voice and data calls in the WLAN (i and j, respectively) and the number of voice calls in the cell (k). The steady-state probability is denoted by $\pi(i, j, k)$. Based on the bandwidth occupancy of voice traffic in the cell and the Weibull distribution $W_b(x, \alpha_d, \beta_d)$ in (1.1) for overall data call size, the mean data call arrival rate to the cell is obtained as

$$\lambda_d^c = \lambda_d \cdot \delta_d^c \cdot \chi_d^c \tag{4.1}$$

$$\delta_d^c = \int_0^{\Omega_d} W_b(x, \alpha_d, \beta_d)dx, \quad \chi_d^c = \sum_{(i,j)} \sum_{k:\, C_d^c(k) \geq R_d^c} \pi(i, j, k)$$

where δ_d^c is the fraction of data calls with a size not greater than Ω_d, $(1 - \chi_d^c)$ is the probability that such a data call is blocked by the cell due to congestion, and $C_d^c(k)$ is the maximum cell capacity available to data traffic when there are k voice calls in progress. Similarly, the mean data call arrival rate to the WLAN is given by

$$\lambda_d^w = \lambda_d \cdot \left[\delta_d^c \cdot (1 - \chi_d^c) + (1 - \delta_d^c) \right] = \lambda_d \cdot \left(1 - \delta_d^c \cdot \chi_d^c \right). \tag{4.2}$$

The corresponding state transition rates of the three-dimensional Markov process are defined in (4.3). Here, N_v^c and N_v^w are the maximum numbers of voice calls admitted in the cell and the WLAN, respectively, $N_d^w(i)$ is the maximum number of data calls allowed in the WLAN with i voice calls in progress,[1] $\xi_d^w(i, j)$ is the mean service rate provided to each data call when there are i voice calls and j data calls in the WLAN, and g_d^w is the mean size of data calls flowing to the WLAN. Note that the transition rate from state (i, j, k) to state $(i - 1, j, k)$ consists of two components. One is due to the completion of the i voice calls in the WLAN with a mean rate of $i \cdot \mu_v$, and the other is due to the completion of the k voice calls in the cell with a mean rate $k \cdot \mu_v$. When one of the k voice calls in the cell completes and makes room for a new voice call, one of the i voice calls in the WLAN can be handed over to the cell. According to the size-based scheme and the overall data call size distribution in (1.1), g_d^w can be derived as

$$
\begin{aligned}
(i, j, k) &\rightarrow (i, j, k+1): \lambda_v, & & i \leq N_v^w,\ j \leq N_d^w(i),\ k \leq N_v^c - 1 \\
(i, j, k) &\rightarrow (i, j, k-1): k \cdot \mu_v, & & i = 0,\ j \leq N_d^w(i),\ 1 \leq k \leq N_v^c \\
(i, j, k) &\rightarrow (i+1, j, k): \lambda_v, & & i \leq N_v^w - 1,\ j \leq N_d^w(i+1),\ k = N_v^c \\
(i, j, k) &\rightarrow (i-1, j, k): (i+k) \cdot \mu_v, & & 1 \leq i \leq N_v^w,\ j \leq N_d^w(i),\ k = N_v^c \\
(i, j, k) &\rightarrow (i, j+1, k): \lambda_d^w, & & i \leq N_v^w,\ 0 \leq j \leq N_d^w(i) - 1,\ k \leq N_v^c \\
(i, j, k) &\rightarrow (i, j-1, k): j \cdot \xi_d^w(i, j)/g_d^w, & & i \leq N_v^w,\ 1 \leq j \leq N_d^w(i),\ k \leq N_v^c
\end{aligned}
\tag{4.3}
$$

$$g_d^w = \frac{(1 - \chi_d^c) \int_0^{\Omega_d} x W_b(x, \alpha_d, \beta_d)dx + \int_{\Omega_d}^{\infty} x W_b(x, \alpha_d, \beta_d)dx}{\delta_d^c \cdot (1 - \chi_d^c) + (1 - \delta_d^c)}. \tag{4.4}$$

The first term in the numerator of (4.4) corresponds to data calls of a size not greater than Ω_d, which are blocked by the cell due to congestion with a probability $(1 - \chi_d^c)$

---

[1] N_v^c, N_v^w, and $N_d^w(i)$ are obtained from the admission regions of the cell and the WLAN, i.e., the feasible sets of vectors (n_v^c, n_d^c) and (n_v^w, n_d^w), respectively. Here, $N_v^c = \max(n_v^c)$, $N_v^w = \max(n_v^w)$, and $N_d^w(i) = \max(n_d^w)$, given $n_v^w = i$.

and overflow to the WLAN. The second term in the numerator accounts for the data calls that have a size larger than Ω_d and are assigned to the WLAN to request admission. The denominator is a normalization constant for the size distribution of data calls flowing to the WLAN.

Due to the interdependence between i and k as shown in the state transition rates of (4.3), the size of the state space does not explode with the third dimension of the Markov process, i.e., the number of voice calls in the cell. The steady-state probabilities $\pi(i, j, k)$ can be obtained by solving a very sparse linear system of balance equations. Then, the voice call blocking probability B_v is given by

$$B_v = \sum_{\substack{(i,j):i \leq N_v^w \\ j > N_d^w(i+1)}} \pi(i, j, N_v^c). \tag{4.5}$$

That is, an incoming voice call is blocked if there are N_v^c voice calls in the cell and not sufficient spare capacity is available for one more voice call, and the WLAN is also congested with i voice calls and j data calls so that, with the j data calls already in progress, the admission of one more voice call in the WLAN will result in delay violation to the admitted i voice calls.

As illustrated in Fig. 4.1, when overload occurs, the mean transfer time under the PS increases dramatically with the offered load and the number of admissible calls (N_d), while the call blocking probability converges and cannot be reduced by increasing N_d. In contrast, in an underload case, the call blocking probability is sufficiently small with a reasonably large value of N_d and the mean transfer time is almost independent of N_d. Similar phenomenon is observed for the SRPT discipline. Hence, the QoS of data calls can be assured by maintaining an underload condition for data traffic in the cell. This can be achieved by properly determining the size threshold Ω_d. Thus, the data call blocking probability B_d is obtained as

$$B_d = \left[\delta_d^c \cdot (1 - \chi_d^c) + (1 - \delta_d^c) \right] B_d^w = (1 - \delta_d^c \cdot \chi_d^c) B_d^w \tag{4.6}$$

where B_d^w is the data call blocking probability of the WLAN, given by

$$B_d^w = \sum_{\substack{(i,j):i \leq N_v^w \\ j+1 > N_d^w(i)}} \sum_{k=0}^{N_v^c} \pi(i, j, k). \tag{4.7}$$

That is, the admission of a new data call should not degrade the WLAN capacity so much that the bandwidth requirement of ongoing voice calls cannot be satisfied. From the Little's law, we have the mean transfer time of data calls served in the WLAN

$$E[T_d^w] = \frac{1}{(1 - B_d^w)\lambda_d^w} \sum_{\substack{(i,j):i \leq N_v^w \\ j \leq N_d^w(i)}} \sum_{k=0}^{N_v^c} j\pi(i, j, k). \tag{4.8}$$

On the other hand, the mean transfer time of data calls admitted to the cell can be obtained with an $M/G/1 - SRPT$ queueing system. This is because data call arrivals to the cell is still a Poisson process with a mean rate λ_d^c given in (4.1). The data call blocking probability is negligibly small if an underload condition is guaranteed by properly selecting the threshold Ω_d. The average bandwidth allocated to data calls is

$$\overline{C}_d^c = \sum_{\substack{(i,j):\, i \leq N_v^w \\ j \leq N_d^w(i)}} \sum_{k=0}^{N_v^c} C_d^c(k)\pi(i, j, k). \tag{4.9}$$

Then, based on the formulas in [40], the mean transfer time is approximated by

$$E[T_d^c] = \int_0^{\Omega_d} \frac{1}{\delta_d^c} W_b(x, \alpha_d, \beta_d)\, \Gamma_d^c(x)dx \tag{4.10}$$

where $\frac{1}{\delta_d^c} W_b(x, \alpha_d, \beta_d)$ $(0 < x \leq \Omega_d)$ is the PDF of the data call size in the cell, and $\Gamma_d^c(x)$ is the conditional transfer time for a data call of size x, given by

$$\Gamma_d^c(x) = \int_0^y \frac{dt}{1 - \rho_d^c(t)} + \frac{\lambda_d^c\left[\int_0^y t^2 g_{L_d}(t)dt + y^2\left(1 - G_{L_d}(y)\right)\right]}{2\left[1 - \rho_d^c(y)\right]^2} \tag{4.11}$$

$$\rho_d^c(y) = \lambda_d^c \int_0^y t g_{L_d}(t)dt, \qquad y = \frac{x}{\overline{C}_d^c} \tag{4.12}$$

$$g_{L_d}(t) = \frac{1}{\delta_d^c} W_b(t, \alpha_d, \beta_d/\overline{C}_d^c), \qquad 0 < t \leq \Omega_d/\overline{C}_d^c. \tag{4.13}$$

Here, $g_{L_d}(\cdot)$ denotes the PDF of a bounded Weibull distribution and $G_{L_d}(\cdot)$ the corresponding CDF. In contrast to the data call size distribution $W_b(x, \alpha_d, \beta_d)$ in (1.1), the scale parameter of $g_{L_d}(\cdot)$ is defined by proportionally modifying β_d with $\overline{C}_d^c$ to switch the unit from data call size to service time.

For comparison purpose, when data calls in the cell are served under the PS discipline, the mean transfer time is approximated by [10]

$$E[T_d^c] = \frac{(\overline{\rho}_d^c)^{N_d^c+1}(N_d^c \, \overline{\rho}_d^c - N_d^c - 1) + \overline{\rho}_d^c}{\lambda_d^c \cdot \left[1 - (\overline{\rho}_d^c)^{N_d^c}\right](1 - \overline{\rho}_d^c)}, \qquad \overline{\rho}_d^c = \rho_d^c(\Omega_d/\overline{C}_d^c) \qquad (4.14)$$

where N_d^c is the maximum number of data calls allowed in the cell, and $\overline{\rho}_d^c$ is the average load factor of data traffic in the cell, which can be obtained from (4.11). Taking into account the size-based access selection for data traffic, the overall average transfer time of data calls can be evaluated by

$$T_d = \frac{\delta_d^c \, \chi_d^c \cdot E[T_d^c] + \left[\delta_d^c \cdot (1 - \chi_d^c) + (1 - \delta_d^c)\right](1 - B_d^w) \cdot E[T_d^w]}{\delta_d^c \, \chi_d^c + \left[\delta_d^c \cdot (1 - \chi_d^c) + (1 - \delta_d^c)\right](1 - B_d^w)}. \qquad (4.15)$$

4.2.2 Determination of Data Size Threshold

In this scheme, voice calls are preferably distributed to the cell for high efficiency and fine QoS. Data traffic should be properly balanced between the two systems correspondingly. Based on the observations in Sect. 4.1, there are some important principles to follow in determining the data size threshold Ω_d.

First, an underload condition should be ensured for data traffic in the cell. That is, the data load factor in the worst case, denoted by $\hat{\rho}_d^c$, is less than 1. Based on (4.11), $\hat{\rho}_d^c$ is evaluated by

$$\hat{\rho}_d^c = \lambda_d^c \int_0^{\Omega_d/R_d^c} t \frac{1}{\delta_d^c} W_b(t, \alpha_d, \beta_d/R_d^c) dt < 1 \qquad (4.16)$$

where R_d^c is the minimum cell bandwidth available to data traffic, and $\frac{1}{\delta_d^c} W_b(t, \alpha_d, \beta_d/R_d^c)$, $0 < t \le \Omega_d/R_d^c$, denotes the PDF of a bounded Weibull distribution with shape parameter α_d and scale parameter β_d/R_d^c. Moreover, data calls with a smaller size usually expect a shorter transfer time than those with a larger size. As data calls in the cell have a smaller size than most of those in the WLAN, our second principle is to guarantee that $E[T_d^c] \le E[T_d^w]$. The mean transfer time $E[T_d^w]$ and $E[T_d^c]$ are given by (4.8) and (4.10), respectively.

Last, when determining the size threshold Ω_d, we should make a good trade-off between user-perceived QoS such as mean data transfer time and call blocking probabilities. An appropriate threshold Ω_d^* can be determined to satisfy the following condition:

$$B_d(\Omega_d) < B_d(\Omega_d^*) \Rightarrow T_d(\Omega_d) > T_d(\Omega_d^*), \quad \forall \, \Omega_d \ne \Omega_d^*. \qquad (4.17)$$

That is, the size threshold Ω_d should be chosen so that the mean transfer time T_d is minimized without increasing the data call blocking probability B_d. As such, the resource utilization is improved without degrading the QoS performance.

To clarify the impact of data call size threshold Ω_d on performance, we carry out some numerical experiments with the following system parameters: the mean voice call arrival rate $\lambda_v = 0.45$ (calls/s), average voice call duration $(\mu_v)^{-1} = 140$ (s), and average data call size $E[L_d] = 64K$ (byte). The parameters for the WLAN and the cell are given in Table 2.1.

Figure 4.2a–c show the impact of data size threshold (Ω_d) on voice and data call blocking probabilities (B_v and B_d, respectively) and mean data transfer time (T_d) in different load conditions (λ_d). It is observed that B_v, B_d, and T_d only slightly decrease with Ω_d when Ω_d is relatively small. After a certain threshold (say, $\Omega_d = 102.4$ kbits), B_v and B_d begin to decrease faster with Ω_d. When Ω_d is sufficiently large (e.g., $\Omega_d \geq 640.0$ kbits), T_d even increases exponentially with Ω_d. The phenomena observed in Fig. 4.2 can be explained as follows. The explosive increase of T_d with a large value of Ω_d is due to congestion in the cell. As seen from (4.2), more data traffic load is assigned to the cell when Ω_d is larger. Due to a small cell bandwidth and high occupancy by voice calls, the data call performance is degraded substantially if the cell is overloaded. On the other hand, when Ω_d is relatively small, the decrease of T_d with Ω_d is attributed to the fact that the cell bandwidth unused by voice traffic can be efficiently utilized by small-size data calls under the SRPT. When Ω_d is sufficiently small to meet the underload condition, the larger the value of Ω_d, the greater the portion of small-size data calls assigned to the cell. Under the SRPT, the small-size data calls in the cell will not stay in the system for such a long duration as in the case where the bandwidth is shared with large-size data calls in a PS manner.

To further evaluate the impact of heavy-tailedness degrees of data call size, we vary the shape parameter α_d ($0 < \alpha_d \leq 1$) in (1.1) and select the scale parameter β_d accordingly to keep the same mean. The smaller the value of Weibull factor α_d, the heavier the tail of the distribution of data call size. Figure 4.3a–c illustrate the voice and data performance versus the size threshold with different degrees of heavy-tailedness. Figure 4.3a, b show that voice and data call blocking probabilities decrease with Ω_d more slowly if α_d is smaller. When Ω_d is relatively small, B_v even increases with a larger α_d. On the other hand, as seen in Fig. 4.3c, the mean data transfer time T_d first slowly decreases with Ω_d until a sufficiently large Ω_d leads to an explosive increase of T_d due to system overload. In contrast to Fig. 4.2c with an exponentially distributed data call size, the reduction of T_d with Ω_d is more evident in the heavy-tailed case. For a smaller α_d (say, 0.2), T_d decreases more slowly and can achieve an even smaller lower bound. This is due to the *"mice-elephants"* property of heavy-tailed distributions. A smaller α_d (i.e., a higher level of heavy-tailedness) implies that there is a larger fraction of even shorter data calls and that less data calls have a much larger size. Given the same size threshold Ω_d, more data calls can then be efficiently served under the SRPT in the cell. As a result, a smaller T_d is achievable with an appropriate size threshold. From Figs. 4.2 and 4.3, we can conclude that the load

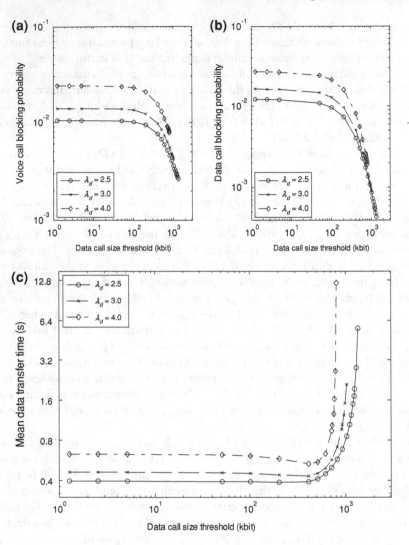

Fig. 4.2 Voice and data call QoS versus data size threshold (Ω_d) with an exponentially distributed data call size ($\alpha_d = 1.0$) and different load conditions of $\lambda_d = 2.5$, 3.0, and 4.0 (calls/s), respectively. **a** Voice blocking probability. **b** Data blocking probability. **c** Mean data transfer time

conditions and traffic characteristics should be properly incorporated in determining the data size threshold.

Taking into account the observations in Figs. 4.2 and 4.3, we present a simple search algorithm in Table 4.1 to determine the data size threshold. Following the principles discussed at the beginning of this section, we apply the Brent's method [32] to find the optimal Ω_d^* that minimizes the mean data transfer time T_d. The constraints, $\hat{\rho}_d^c < 1$ and $T_d^c \leq T_d^w$, are incorporated in the Brent's method by setting $T_d = \infty$ if

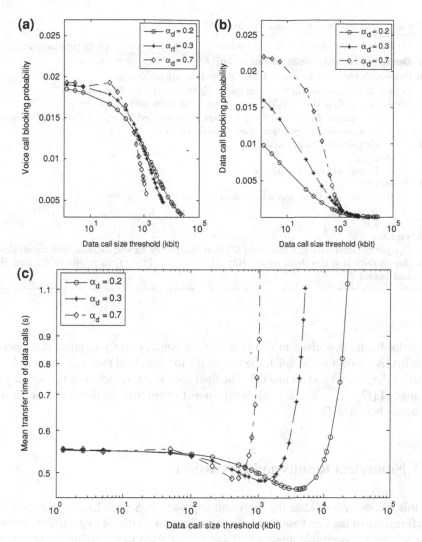

Fig. 4.3 Voice and data call QoS versus data size threshold (Ω_d) with mean data call arrival rate $\lambda_d = 3.6$ (calls/s) and different heavy-tailedness of data call size, i.e., $\alpha_d = 0.2$, 0.3, and 0.7, respectively. **a** Voice blocking probability. **b** Data blocking probability. **c** Mean data transfer time

these constraints are violated. As a superlinear search method, the Brent's method can efficiently locate the minimum. In each iteration, the QoS metrics are evaluated only once with a given trial size threshold. The analytical approach in Sect. 4.2.1 is employed to effectively evaluate the QoS metrics such as B_v, B_d, and T_d. Hence, the running overhead for determining the size threshold should be affordable and Ω_d can be adapted to traffic load variations. Moreover, it is observed in Figs. 4.2c and 4.3c that T_d may be sensitive to Ω_d in the neighborhood of Ω_d^*. Therefore, the

Table 4.1 Search algorithm for data size threshold

1: Derive WLAN capacity region of vectors (n_v^w, n_d^w) to meet stability constraints
2: Derive cell capacity region of vectors (n_v^c, n_d^c) to satisfy $\frac{E_b}{N_0}$ requirements
3: Set search range for data size threshold as $[\Omega_{\{d,min\}}, \Omega_{\{d,max\}}]$
   // Search for optimal Ω_d that minimizes T_d by Brent's method [32]
4: **for** $i = 1, ..., N_{iter}$ **do**   // Try N_{iter} rounds of iterations at maximum
       // Constraints, $\hat{\rho}_d^c < \varepsilon$ and $T_d^c \le T_d^w$, are incorporated by setting $T_d = \infty$ if violated
5:     **if** A parabolic interpolation is acceptable **then**
6:         Construct trial parabolic fits
7:     **else**
8:         Resort to golden section search
9:     **end if**
10:    **if** Desired precision is reached **then**
11:        break
12:    **end if**
13: **end for**
14: Output optimal data size threshold Ω_d^* that minimizes T_d and satisfies QoS constraints
15: Adapt data size threshold within $[\Omega_d^* \cdot (1 - \tau), \ \Omega_d^* \cdot (1 + \tau)]$ to minimize B_v and B_d
    and bound T_d

---

underload condition given in (4.16) is applied conservatively to guarantee system stability. As shown in Table 4.1, the bound for the data load factor $\hat{\rho}_d^c$ is set to be ε, which is less than 1 and around 0.9. The final data size threshold can further vary in a range of $[\Omega_d^* \cdot (1 - \tau), \ \Omega_d^* \cdot (1 + \tau)]$, so as to minimize B_v and B_d while ensure an upper-bounded T_d.

4.3 Numerical Results and Discussion

In this section, we validate the analytical approach in Sect. 4.2.1, and compare the performance of the size-based load sharing scheme with that of two other interworking schemes. System parameters in Table 2.1 are used in the following numerical analysis. For investigation simplicity, we consider constant voice traffic load with $\lambda_v = 0.45$ (calls/s).

4.3.1 Accuracy Validation

Figure 4.4 shows the analytical and simulation results of voice and data call blocking probabilities (B_v and B_d, respectively) and mean transfer time of data calls (T_d) when the data call size is exponentially distributed, i.e., the Weibull factor $\alpha_d = 1$. It can be seen that the analytical results match well the simulation results under

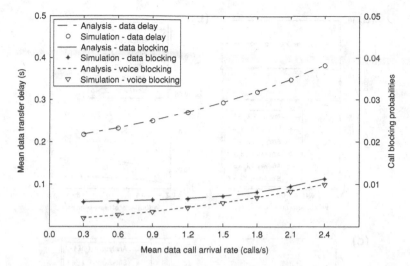

Fig. 4.4 Analytical and simulation results of voice and data call blocking probabilities (B_v and B_d, respectively) and mean data transfer time (T_d) versus mean data call arrival rate (λ_d) with an exponentially distributed data call size ($\alpha_d = 1.0$)

different load conditions (λ_d). Figure 4.5a–c further illustrate the cases with a heavy-tailed data call size, i.e., $0 < \alpha_d < 1$. Similarly, the analytical results agree with the simulation results, except that the voice and data call blocking probabilities are slightly overestimated when $\alpha_d \leq 0.3$. This is due to the increase of heavy-tailedness with a small α_d. In our analytical model in Sect. 4.2.1, we assume that the QoS of data calls in the WLAN is insensitive to the data call size distribution under the PS service discipline. Due to varying WLAN capacity, the insensitivity is impaired and the call-level QoS may improve when a greater variability is induced with the heavy-tailed call size [23]. Nonetheless, the insensitivity is expected to retain when the call blocking probabilities are sufficiently small. As seen in Fig. 4.5, with a relatively light traffic load and a smaller data call blocking probability, the gap between the analytical results and simulation results is much smaller when $\alpha_d \leq 0.3$. As the system is usually designed to ensure call blocking probabilities in the order of 10^{-3}–10^{-2}, the analytical model in Sect. 4.2.1 is valid for performance analysis.

4.3.2 Performance Improvement

To evaluate the effectiveness of the size-based load sharing scheme, we compare its performance with that of two other interworking schemes, i.e., the admission control scheme with randomized access selection in Chap. 3 and the service-differentiated resource allocation scheme proposed in [46]. The service-differentiated scheme extends the WLAN-first resource allocation scheme with service differentiation.

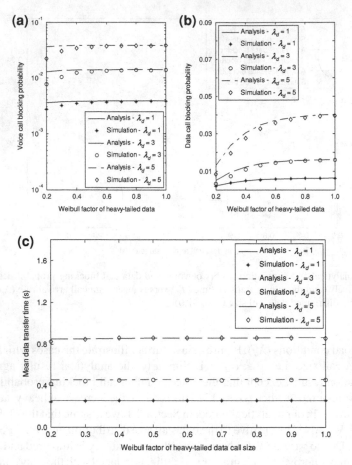

Fig. 4.5 Analytical and simulation results of voice and data call QoS versus Weibull factor (α_d) for a heavy-tailed data call size with $\lambda_d = 1$, 3, and 5 (calls/s), respectively. **a** Voice blocking probability. **b** Data blocking probability. **c** Mean data transfer time

A new voice call originating in the double-coverage area first attempts to get admission to the cell, whereas new data calls in the double-coverage area first request the WLAN for admission. When a mobile moves from the cellular-only area into the WLAN coverage, its associated voice calls are not handed over from the cell to the WLAN. This is to avoid QoS degradation induced by vertical hand-off and the inefficient real-time service support of the WLAN. In contrast, ongoing data calls served by the cell will attempt to hand over to the WLAN for the larger bandwidth.

Figure 4.6a–c show the performance of the three schemes in terms of voice and data call blocking probabilities (B_v and B_d, respectively) and mean data transfer time (T_d), respectively. Significant performance improvement is observed with the size-based scheme. For example, in the case of $\lambda_d = 3.6$ (calls/s), B_v of the size-based

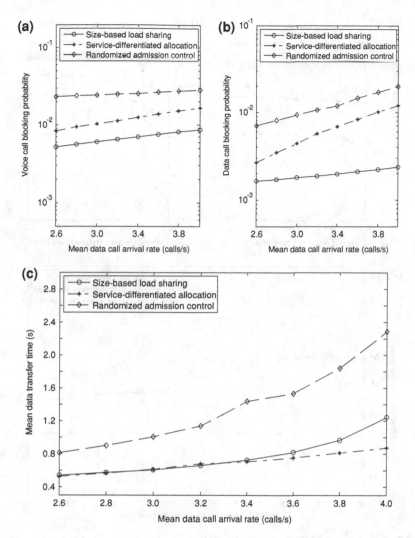

Fig. 4.6 Performance of different interworking schemes versus mean data call arrival rate (λ_d) with an exponentially distributed data call size ($\alpha_d = 1.0$). **a** Voice blocking probability. **b** Data blocking probability. **c** Mean data transfer time

scheme is 71.4 % smaller than that of the randomized scheme, while B_d is reduced by more than 85.6 % and T_d is 46.8 % lower. A performance gain of 45.4 % and 74.8 % is achieved by the size-based scheme with respect to the service-differentiated scheme for B_v and B_d, respectively, although T_d of the two schemes is very close. In some cases, T_d of the service-differentiated scheme is even slightly lower than that of the size-based scheme. However, this low mean data transfer time of the service-differentiated scheme is achieved at the expense of much higher call blocking

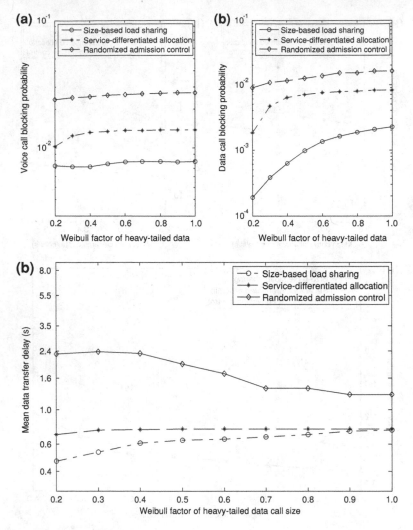

Fig. 4.7 Performance of different interworking schemes under a heavy-tailed data call size versus Weibull factor (α_d) with mean data call arrival rate $\lambda_d = 3.6$ (calls/s). **a** Voice blocking probability. **b** Data blocking probability. **c** Mean data transfer time

probabilities B_v and B_d. The size-based load sharing scheme still outperforms the other two schemes.

Figure 4.7a–c show the performance of the three schemes with different Weibull factors α_d, i.e., different degrees of heavy-tailedness for the data call size. As seen, an even larger performance gain is achievable with the size-based scheme for B_d and T_d when α_d is smaller, i.e., the data call size is distributed with a heavier tail. For example, when $\alpha_d = 0.2$, B_d of the size-based scheme is more than 95 % smaller than that of the other two schemes, while there is a reduction around 87.7 % when

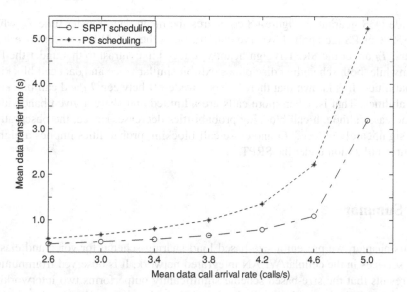

Fig. 4.8 Mean transfer time (T_d) of data calls with $\alpha_d = 1.0$ under SRPT or PS scheduling disciplines in the cell

$\alpha_d = 0.8$. Similarly, when α_d decreases from 0.8 to 0.2, the reduction of T_d with respect to the randomized scheme increases from 49.6 % to 79.7 %. In comparison with the service-differentiated scheme, the size-based scheme reduces T_d by 7.7 % when $\alpha_d = 0.8$ and by 32.8 % when $\alpha_d = 0.2$. A larger reduction of T_d with respect to α_d is due to the much higher call blocking probabilities, which restrict the total admissible traffic load to share the bandwidth.

The significant performance gain observed in Figs. 4.6 and 4.7 lies in the fact that the size-based load sharing scheme not only takes advantage of the complementary QoS of the integrated systems in initial call assignment, but also exploits vertical handoff in call reassignment to maximize the multiplexing gain. Moreover, the data size threshold can be appropriately determined with the approach given in Sect. 4.2.1, which effectively takes into account the load conditions and heavy-tailedness of data call size. Nonetheless, the size-based scheme requires that the data call size be known a priori via session signaling. The signaling and control overhead for dynamic vertical handoff may increase the implementation complexity.

4.3.3 Overload Protection via SRPT Scheduling

As discussed in Sect. 4.1, data calls in the cell are served under the SRPT, which can be enabled by the centralized resource allocation and benefit the system with the best performance achievable. The advantage of SRPT is particularly more evident in system overload when it is more challenging for the cell of small bandwidth to provide QoS guarantee. Figure 4.8 compares the mean data transfer time T_d when the SRPT or PS are applied to serve data traffic in the cell. When system overload occurs, T_d under the SRPT is significantly reduced in contrast to that under the PS. Meanwhile, both scheduling disciplines exhibit similar voice and data call blocking probabilities. It is known that there exists a trade-off between T_d and call blocking probabilities. That is, when more calls are admitted and share a given bandwidth, T_d increases although call blocking probabilities decrease. Hence, the observation of a significantly reduced T_d and close call blocking probabilities implies a higher resource utilization under the SRPT.

4.4 Summary

In this chapter, we present a size-based load sharing scheme for voice and elastic data services in the cellular/WLAN integrated network. It is observed from numerical results that the size-based scheme significantly outperforms two interworking schemes with randomized access selection and service-differentiated resource allocation. The main features and observations are as follows:

- In the size-based load sharing scheme, voice traffic load is preferably admitted into the cell by means of both initial call admission and access reselection via dynamic vertical handoff. A large multiplexing gain is achieved by pooling the free bandwidth in the two systems to effectively serve elastic data traffic.
- The heavy-tailedness of data call size is exploited by an access selection strategy based on a size threshold. Further, the system capacity is improved by using the efficient SRPT scheduling for data calls in the cell.
- The system performance is evaluated accurately with an analytical approach. It characterizes the heavy-tailedness of data traffic and dynamic vertical handoff triggered by network states. The data call size threshold can be determined with the analytical model.

Chapter 5
Conclusions and Future Directions

In this chapter, we summarize the main observations of the interworking solutions and highlight future research directions.

5.1 Research Conclusions

The objective of this brief is to present the state-of-the-art solutions to effective cellular/WLAN interworking and efficient utilization of overall resources. Specifically, we obtain the following important insights on cellular/WLAN interworking:

- *Effective interworking needs to take good advantage of the two-tier overlay structure and complementary strengths of cellular networks and WLANs.* The centralized control of the cellular network enables dedicated resource allocation and hard QoS guarantee. The large cell size and ubiquitous coverage reduces handoff frequency. In contrast, The WLAN provides a much larger bandwidth at a lower cost, but suffers from contention-based access overhead and disjoint small coverage. Such network characteristics must be considered and captured in the interworking schemes.
- *The unique characteristics of multi-service traffic and user mobility are essential to achieve a high overall resource utilization.* Different services require different QoS deliveries, while the cellular network and WLANs provide complementary support for real-time voice service and elastic data services. The heavy-tailed property of data traffic can be addressed to improve resource utilization. Further, the location-dependent mobility model characterizes the low mobility within WLAN-covered indoor hotspots. Accordingly, interworking schemes can better exploit the large WLAN bandwidth whenever available.
- *Interworking efficiency can be optimized by jointly considering access selection, call admission control, and load sharing.* Access selection matches a service request with a preferable network support, while call admission control limits the accepted traffic load and ensure QoS guarantee. Network dynamics can be

W. Song and W. Zhuang, *Interworking of Wireless LANs and Cellular Networks*,   61
SpringerBriefs in Computer Science, DOI: 10.1007/978-1-4614-4379-7_5,
© The Author(s) 2012

further exploited via vertical handoff to enhance load sharing. As such, the multi-service traffic load can be properly distributed to the integrated cell and WLAN and achieve a large multiplexing gain.

5.2 Future Extensions

Although the complementary strength of the cellular network and WLAN has promoted their interworking, the network heterogeneity poses many research challenges to the interworking. The solutions reviewed in this brief leverage access selection and call admission control to take advantage of the cellular/WLAN interworking for load sharing. There are many open issues to extend the research in the following directions:

- *Video streaming can be further considered in the interworking schemes in addition to voice and data services.* Due to video coding and compression, video traffic exhibits high burstiness and variation. Particularly, flow continuity and playout smoothness are essential for streaming services. New challenges are introduced to ensure high-quality video streaming over the cellular/WLAN integrated network. The video flow information such as clip length and encoding rate can be derived from control signaling and considered in access selection and admission control. Adaptive coding with fine-granular scalability can be exploited to minimize playback interruption [44].
- *It is even more challenging to implement the cellular/WLAN interworking for mobile hotspots in a vehicular environment.* Most of the previous studies on the cellular/WLAN interworking focus on a simple scenario with static WLAN deployment in an indoor environment such as offices, hotels, and airport terminals. In contrast with the static indoor WLANs, there are ever-increasing demands for systematic deployment of mobile hotspots [18], which are usually in and around a moving vehicle, such as a bus, a railway train, and even a flight cabin. For mobile hotspots in a vehicular environment, it is usually not feasible to take advantage of the overlay structure as discussed in this brief. The data traffic from the mobile hotspot is often multiplexed at a mobile WLAN router and then relayed to a cellular base station [45]. Due to the heavy-tailedness of data file size, the aggregate traffic exhibits self-similarity and poses more stringent constraints for interworking efficiency.
- *The cellular/WLAN interworking can be extended to a multi-hop relay scenario with an increased level of decentralization.* Mobile multi-hop relay (MMR) is drafted in IEEE 802.16j for worldwide inter-operability for microwave access (WiMAX). Also, IEEE 802.11s is specified to amend WLAN with mesh networking. In fact, multi-hop relay is even more favorable for the cellular/WLAN interworking. As WLANs are usually deployed in disjoint hotspots, the relay between feasible access points can eliminate the necessity of single-hop connection to the cellular base station. Legacy single-mode mobile terminals can

also benefit from the relay of dual-mode mobile terminals [57]. However, such a multi-hop infrastructure and distributed control introduce more challenges for the cellular/WLAN interworking.

References

1. 3GPP: Selection procedures for the choice of radio transmission technologies of the UMTS. 3GPP TS 30.03 V3.2.0 (1998)
2. 3GPP: Quality of service (QoS) concept and architecture. 3GPP TS 23.107 V7.0.0 (2007)
3. 3GPP: Services and service capabilities. 3GPP TS 22.105 V8.4.0 (2007)
4. Banerjee, N., Wu, W., Das, S.K.: Mobility support in wireless Internet. IEEE Wirel. Commun. Mag. 10(5), 54–61 (2003)
5. Bansal, N., Harchol-Balter, M.: Analysis of SRPT scheduling: investigating unfairness. ACM SIGMETRICS Perform. Eval. Rev. 29(1), 279–290 (2001)
6. Bari, F., Leung, V.: Automated network selection in a heterogeneous wireless network environment. IEEE Netw. 21(1), 34–40 (2007)
7. Benameur, N., Fredj, S.B., Delcoigne, F., Oueslati-Boulahia, S., Roberts, J.W.: Integrated admission control for streaming and elastic traffic. In: Proceedings of 2nd International Workshop on Quality of Future Internet Services, pp. 69–81 (2001)
8. Boxma, O.J., Gabor, A.F., Nunez-Queija, R., Tan, H.P.: Performance analysis of admission control for integrated services with minimum rate guarantees. In: Proceedings of 2nd Conference on Next Generation Internet Design and Engineering (NGI), pp. 41–47 (2006)
9. Crovella, M.E., Bestavros, A.: Self-similarity in World Wide Web traffic: evidence and possible causes. IEEE/ACM Trans. Networking 5(6), 835–846 (1997)
10. Delcoigne, F., Proutière, A., Régnié, G.: Modeling integration of streaming and data traffic. Perform. Eval. 55(3–4), 185–209 (2004)
11. Fang, Y., Zhang, Y.: Call admission control schemes and performance analysis in wireless mobile networks. IEEE Trans. Veh. Technol. 51(2), 371–382 (2002)
12. Feldmann, A., Whitt, W.: Fitting mixtures of exponentials to long-tail distributions to analyze network performance models. Perform. Eval. 31(3–4), 245–279 (1998)
13. Fredj, S.B., Bonald, T., Proutière, A., Régnié, G., Roberts, J.W.: Statistical bandwidth sharing: a study of congestion at flow level. In: Proceedings of ACM SIGCOMM, pp. 111–122 (2001)
14. Fredj, S.B., Oueslati-Boulahia, S., Roberts, J.W.: Measurement-based admission control for elastic traffic. In: Proceedings of 17th International Teletraffic Congress, pp. 161–172 (2001)
15. Furuskär, A., Zander, J.: Multiservice allocation for multiaccess wireless systems. IEEE Trans. Wirel. Commun. 4(1), 174–184 (2005)
16. Garcia-Martin, M., Isomaki, M., Camarillo, G., Loreto, S.: A session description protocol (SDP) offer/answer mechanism to enable file transfer. Internet Draft (2007)
17. Gelabert, X., Peréz-Romero, J., Sallent, O., Agustí, R.: A Markovian approach to radio access technology selection in heterogeneous multiaccess/multiservice wireless networks. IEEE Trans. Mobile Comput. 7(10), 1257–1270 (2008)

W. Song and W. Zhuang, *Interworking of Wireless LANs and Cellular Networks*,
SpringerBriefs in Computer Science, DOI: 10.1007/978-1-4614-4379-7,
© The Author(s) 2012

18. Hasan, M.M., Mark, J.W., Shen, X.: A link adaptation scheme for the downlink of mobile hotspot. Wirel. Commun. Mobile Comput. (2011)
19. Huber, J.F., Weiler, D., Brand, H.: UMTS, the mobile multimedia vision for IMT-2000: a focus on standardization. IEEE Commun. Mag. **38**(9), 129–136 (2000)
20. Kelly, F.P.: Reversibility and Stochastic Networks. Wiley, New York (1979)
21. Kleinrock, L.: Queueing Systems, Volume 1: Theory. John Wiley and Sons, New York (1975)
22. Lincke-Salecke, S.: Load shared integrated networks. In: Proceedings of 5th European Personal Mobile Communications Conference (EPMCC), pp. 225–229 (2003)
23. Litjens, R., Boucherie, R.J.: Elastic calls in an integrated services network: the greater the call size variability the better the QoS. Perform. Eval. **52**(4), 193–220 (2003)
24. Lundevall, M., Olin, B., Olsson, J., Wiberg, N., Wanstedt, S., Eriksson, J., Eng, F.: Streaming applications over HSDPA in mixed service scenarios. In: Proceedings IEEE VTC Fall, vol. 2, pp. 841–845 (2004)
25. Ma, L., Yu, F., Leung, V.C.M.: Performance improvements of mobile SCTP in integrated heterogeneous wireless networks. IEEE Trans. Wirel. Commun. **6**(10), 3567–3577 (2007)
26. Maheshwari, K., Kumar, A.: Performance analysis of microcellization for supporting two mobility classes in cellular wireless networks. IEEE Trans. Veh. Technol. **49**(2), 321–333 (2000)
27. Marsan, M.: Performance analysis of hierarchical cellular networks with generally distributed call holding times and dwell times. IEEE Trans. Wirel. Commun. **3**(1), 248–257 (2004)
28. Montes, W., Gomez, G., Cuny, R., Paris, J.F.: Deployment of IP multimedia streaming services in third-generation mobile networks. IEEE Wirel. Commun. Mag. **9**(5), 84–92 (2002)
29. Naghshineh, M., Acampora, A.: QoS provisioning in micro-cellular networks supporting multiple classes of traffic. Wirel. Netw. **2**(3), 195–203 (1996)
30. Niyato, D., Hossain, E.: A noncooperative game-theoretic framework for radio resource management in 4G heterogeneous wireless access networks. IEEE Trans. Mobile Comput. **7**(3), 332–345 (2008)
31. Park, H.S., Yoon, S.H., Kim, T.H., Park, J.S., Do, M.S., Lee, J.Y.: Vertical handoff procedure and algorithm between IEEE 802.11 WLAN and CDMA cellular network. In: Proceedings of 7th CDMA International Conference on Mobile, Communications (2002)
32. Press, W.H., Teukolsky, S.A., Vetterling, W.T., Flannery, B.P.: Numerical Recipes in C, The Art of Scientific Computing, 2nd edn. Cambridge University Press, Cambridge (1999)
33. Queija, R.N.: Processor-sharing models for integrated-services networks. Ph.D. thesis, Eindhoven University of Technology (2000)
34. Ramjee, R., Towsley, D., Nagarajan, R.: On optimal call admission control in cellular networks. Wirel. Netw. **3**(1), 29–41 (1997)
35. Rezaul, K.M., Pakštas, A.: Web traffic analysis based on EDF statistics. In: Proceedings 7th Annual PostGraduate Symposium on the Convergence of Telecommunications, Networking and Broadcasting (PGNet) (2006)
36. Roberts, J.W., Massoulié, L.: Bandwidth sharing and admission control for elastic traffic. ITC Specialist, Seminar (1998)
37. Rodrigues, E.B., Olsson, J.: Admission control for streaming services over HSDPA. In: Proceedings of AICT/SAPIR/ELETE, vol. 00, pp. 255–260 (2005)
38. Ross, K.W.: Multiservice Loss Models for Broadband Telecommunication Networks. Springer-Verlag, Berlin (1995)
39. Salkintzis, A.K.: Interworking techniques and architectures for WLAN/3G integration toward 4G mobile data networks. IEEE Wirel. Commun. Mag. **11**(3), 50–61 (2004)
40. Schrage, L.E., Miller, L.W.: The queue M/G/1 with the shortest remaining processing time discipline. Oper. Res. **14**(4), 670–684 (1966)
41. Si, P., Ji, H., Yu, F.R.: Optimal network selection in heterogeneous wireless multimedia networks. ACM/Springer Wirel. Netw. **16**(5), 1277–1288 (2010)

42. Song, Q., Jamalipour, A.: Network selection in an integrated wireless LAN and UMTS environment using mathematical modeling and computing techniques. IEEE Wirel. Commun. Mag. **12**(3), 42–48 (2005)
43. Song, W., Jiang, H., Zhuang, W.: Performance analysis of the WLAN-first scheme in cellular/WLAN interworking. IEEE Trans. Wirel. Commun. **6**(5), 1932–1952 (2007)
44. Song, W., Zhuang, W.: Resource allocation for conversational, streaming, and interactive services in cellular/WLAN interworking. In: Proceedings of the IEEE GLOBECOM (2007)
45. Song, W., Zhuang, W.: Resource reservation for self-similar data traffic in cellular/WLAN integrated mobile hotspots. In: Proceedings of the IEEE ICC (2010)
46. Song, W., Zhuang, W., Cheng, Y.: Load balancing for cellular/WLAN integrated networks. IEEE Netw. **21**(1), 27–33 (2007)
47. Stemm, M., Katz, R.H.: Vertical handoffs in wireless overlay networks. Mobile Netw. Appl. **3**(4), 335–350 (1998)
48. Stevens-Navarro, E., Lin, Y., Wong, V.W.: An MDP-based vertical handoff decision algorithm for heterogeneous wireless networks. IEEE Trans. Veh. Technol. **57**(2), 1243–1258 (2008)
49. Stevens-Navarro, E., Mohsenian-Rad, A.H., Wong, V.W.: Connection admission control for multi-service integrated cellular/WLAN system. IEEE Trans. Veh. Technol. **57**(6), 3789–3800 (2008)
50. Stevens-Navarro, E., Wong, V.W.: Comparison between vertical handoff decision algorithms for heterogeneous wireless networks. In: Proceedings of IEEE VTC, Spring, vol. 2, pp. 947–951 (2006)
51. Stevens-Navarro, E., Wong, V.W., Lin, Y.: A vertical handoff decision algorithm for heterogeneous wireless networks. In: Proceedings of the IEEE WCNC (2007)
52. Sun, C., Stevens-Navarro, E., Shah-Mansouri, V., Wong, V.W.: A constrained MDP-based vertical handoff decision algorithm for 4G heterogeneous wireless networks. ACM/Springer Wirel. Netw. **17**(4), 1063–1081 (2011)
53. Thajchayapong, S., Peha, J.: Mobility patterns in microcellular wireless networks. IEEE Trans. Mobile Comput. **5**(1), 52–63 (2006)
54. Tijms, H.C.: Stochastic Models—An Algorithm Approach. Wiley, New York (1994)
55. Tran-Gia, P., Hubner, F.: An analysis of trunk reservation and grade of service balancing mechanisms in multiservice broadband networks. In: Proceedigns of IFIP, Workshop TC6, pp. 83–97 (1993)
56. Vanem, E., Svaet, S., Paint, F.: Effects of multiple access alternatives in heterogeneous wireless networks. In: Proceedings of the IEEE WCNC, vol. 3, pp. 1696–1700 (2003)
57. Wei, H.Y., Gitlin, R.D.: Two-hop-relay architecture for next-generation WWAN/WLAN integration. IEEE Wirel. Commun. Mag. **11**(2), 24–30 (2004)
58. Wolff, R.W.: Stochastic Modeling and the Theory of Queues. Prentice Hall, Englewood Cliffs (1989)
59. Yoon, K., Hwang, C.: Multiple Attribute Decision Making: An Introduction. Sage Publications, London (1995)
60. Yu, F., Krishnamurthy, V.: Optimal joint session admission control in integrated WLAN and CDMA cellular network. IEEE Trans. Mobile Comput. **6**(1), 126–139 (2007)
61. Zhang, J., Yu, F., Wang, X., Chan, H., Leung, V.: SIP Handbook: Services, Technologies, and Security. SIP and vertical handoffs in heterogeneous wireless networks, pp. 253–276. CRC Press (2008)
62. Zhang, W.: Handover decision using fuzzy MADM in heterogeneous networks. In: Proceedings IEEE WCNC, vol. 2, pp. 653–658 (2004)